Oil on Troubled Waters

Oil on Troubled Waters

Gulf Wars, 1980–91

John Creighton

Echoes

British Library Cataloguing-in-Publication Data
Creighton, John
Oil on troubled waters : Gulf Wars, 1980–91.
I. Title
955.054

ISBN 1-87339-545-0

First published 1992 by
Echoes, 26 Westbourne Grove,
London W2 5RH

Typeset by Contour Typesetters, Southall, London
Printed and bound in Great Britain by
Redwood Press Limited, Melksham, Wiltshire

Contents

Abbreviations

ABC	American Broadcasting Company
AFP	Agence France Presse
AS	air/surface missile
AWACS	Airborne Warning And Control System
BBC	British Broadcasting Corporation
bbl	barrel(s)
b/d	barrel(s)/day
BIS	Bank for International Settlements
BP	British Petroleum
CPEs	centrally planned economies
CTL	constructive total loss
EC	European Community
FIS	Front Islamique du Salut (Algeria)
GCC	Gulf Co-operation Council
ICO	Islamic Conference Organization
ICP	Iraqi Communist Party
IEA	International Energy Agency
IISS	International Institute for Strategic Studies
IMF	International Monetary Fund
INOC	Iraq National Oil Company
IPC	Iraq Petroleum Company
IPSA-1	Iraq Pipeline Saudi Arabia No 1
IPSA-2	Iraq Pipeline Saudi Arabia No 2
IRP	Islamic Revolutionary Party (Iran)
KDP	Kurdish Democratic Party (Iraq)
KDPI	Kurdish Democratic Party of Iran
KOTC	Kuwait Oil Tanker Company
MEES	*Middle East Economic Survey*
MENAS	Middle East Navigation Aids Service
NAM	Non-Aligned Movement
NATO	North Atlantic Treaty Organization
NIOC	National Iranian Oil Company
NITC	National Iranian Tanker Company
OAPEC	Organization of Arab Petroleum Exporting Countries

OECD	Organization for Economic Co-operation and Development
OMR	*Oil Market Report*
OPEC	Organization of Petroleum Exporting Countries
PKK	Kurdish Workers' Party (Turkey)
PLO	Palestine Liberation Organization
PUK	Patriotic Union of Kurdestan (Iraq)
RAF	Royal Air Force (Britain)
RDF	Rapid Deployment Force
SA	surface/air missile
SAIRI	Supreme Assembly of the Islamic Revolution in Iraq
SAS	Special Air Service Regiment (Britain)
SBM	single-buoy mooring
SOMO	State Oil Marketing Organization (Iraq)
SPR	Strategic Petroleum Reserve (US)
SS	surface/surface missile
UAE	United Arab Emirates
UAR	United Arab Republic
ULCC	Ultra Large Crude Carrier
UN	United Nations
UNIKOM	United Nations Iraq-Kuwait Observation Mission
US	United States
USSR	Union of Soviet Socialist Republics
VLCC	Very Large Crude Carrier

Preface

This book is being completed at the end of 1991. The original idea was to confine it to the Iran-Iraq war, with which it is still mainly concerned and which in many ways came to be dwarfed by the crisis over Kuwait which began almost exactly as the first war was finally approaching its end. The first 'Gulf war' had gone on for so long that many of us may have begun to wonder whether it was not likely to continue in some way or other until the end of the century—or perhaps into another. It had after all lasted so much longer than the Second World War. Few countries were wholly unaffected by it, if only as consumers of oil, and its continuing as long as it did was a source of dismay to much, though perhaps not quite all, of the international community. The idea of writing its story first came to me a year or so after being compelled by illness to give up my work in Baghdad in 1983. I decided to try to bring out not only the development of events and the historical background but also the significance of oil, the dominant element in the economy of the entire region.

In assembling the picture, I have received more help than I had dared hope for, not simply from old colleagues in the Western oil industry and some traditional mines of information in the field of international affairs—the British Library, the Public Record Office, the libraries of the Foreign and Commonwealth Office, the Bank of England, Cambridge University's Middle East Centre, London University's School of Oriental and African Studies (SOAS), the London bureau of the United Nations, the Royal United Services Institute and the International Institute for Strategic Studies—but also the BBC's Monitoring and Arabic Services and the treasure-chest of the daily shipping newspaper, *Lloyd's List*. Going back a little farther, I owe an immense debt for the experience and insight gained from colleagues in the oil industries of three particular Middle East countries—Iran, Iraq and Saudi Arabia. I am also indebted to the *Middle East Economic Survey*, Frank Cass Publishers and the BBC Monitoring Service for permission to quote from their publications. I have supplemented this material with what I have gained by reading other periodicals and from personal notes and recollection.

I have received much-needed encouragement in my endeavours from my former tutor at Cambridge, Professor William Brock, and Sir James

Craig who, in addition to supervising my Arabic studies at Shemlan some thirty years ago, gave me valuable counsel over the learning of Persian before I set off for Tehran a few years later. Dr Michael Burrell at SOAS was an ever-flowing fountain of advice on sources and reasons for persevering with a task that at times seemed likely to last even longer than the fighting itself. My family have kept my word-processor in working order and cheerfully endured the domination of their home by the remarkable quantities of paper generated by the project. My editor, Jana Gough, approached her task with exceptional fortitude.

Colleagues both in the oil industry and elsewhere have examined the first draft and, grateful as I am for all their comments, it would be invidious to single any out for special mention.

London, August 1991

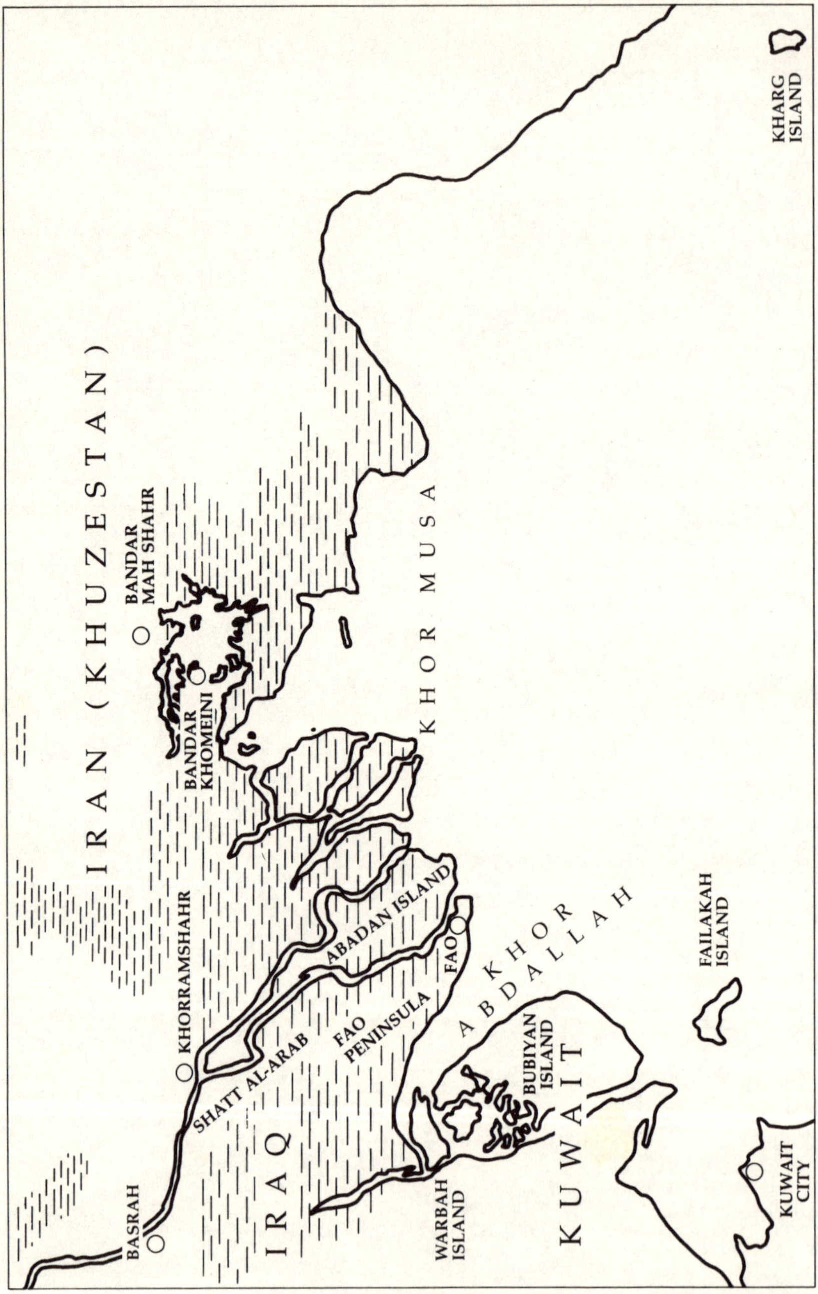

Map I. The Shatt al-Arab, 1980

1
The Outbreak of War, 1980

Although this is the story of a war I had every reason to treat seriously while I was close to it, I offer no excuse for the fact that the account begins on a trivial note. Like many an evening television documentary, or even melodrama, it does so with the buzz and clatter of a news agency teleprinter. The date is 23 September 1980 and the foreign news correspondents covering the affairs of the Middle East on the spot, from such vantage points as the bars of Beirut's few still-operational hotels, received a jolt from messages that the agency machines were beginning to churn out reports[1] of further escalation the previous day of the long-running border dispute[2] between Iraq and Iran. These reports sent them bustling off to organize air tickets to the head of the Gulf, designating it Persian or Arabian according to the nationality of the airline or the destination selected, to check out the situation and track down flesh with which to clothe the bare bones.

There were three main reports to be pursued. The first was of an Iraqi incursion in unprecedented strength into Iranian territory across the Shatt al-Arab[3] near Basrah. The second spoke of Iraq's having attacked Tehran's Mehrabad (both civil and military) airport: Iran had retaliated with air attacks on Baghdad[4] and on a petrochemical complex near

Basrah.[5] The third report, from the Gulf navigation service MENAS, was of Iran's having declared the half of the Gulf adjacent to its coastline a war zone and ordering merchant shipping to keep out.

It duly emerged that the Iraqi incursion amounted to a four-pronged attack,[6] partly in the south near Basrah and partly farther north along the frontier, to the east of Baghdad, in what later evolved into the 'central front'. The attack in the south was focused in the main on Dezful, where Iran had a major air base, and on Ahwaz, the principal town of Iran's southwestern province of Khuzestan and site of the field headquarters of the country's inland oil-production operations, with an important oilfield on the spot and the junction of the crude oil pipeline system leading northwards to domestic refining centres. Another objective was the area containing the port of Khorramshahr and the oil refinery, town and airport of Abadan.

In the push across the frontier in the centre, the Iraqis drove towards two specific areas: Qasr-e-Shirin and nearby Sar-e-Pol-e-Zahab on the road to Tehran; and Mehran and Bostan/Susangerd, farther south and guarding the approach to central Khuzestan.

One account[7] reported 'Arab diplomats' as saying that Iraq had used a strike force of 10,000 to gain 90 square miles (230 sq. km). Others varied, indicating in one case as many as nine Iraqi divisions deployed.[8] What is certain is that the Iraqis immediately took Qasr-e-Shirin,[9] also inflicting severe damage on the Abadan refinery[10] and beginning a siege of Khorramshahr which they later claimed, a little prematurely, to have captured.

What began as something of a *blitzkrieg*, intended to paralyse Iran's military and oil-transmission systems and the political leadership's control of the country, soon slowed down,[11] encountering a totally unexpected[12] capacity to resist on the part of the Iranian forces.[13] The *Sunday Times* summarized it in these terms: 'Iraqi whirlwind campaign ends in war of desert trenches.'[14]

Quite early on, Iraq made known a number of demands to which Iran would have to accede if the crisis were to be brought to an end.[15] These included the return to Sharjah and Ras al-Khaimah of the islands Abu Musa and the two Tunbs, occupied by Iran in 1971;[16] a modification of the 1975 Algiers agreement which Iraq had unilaterally abrogated at the beginning of the war;[17] autonomy for the Arab, Kurdish and Baluchi minorities in Iran; and the 'acceptance of certain conditions' (no doubt concerning non-interference in Iraq's domestic affairs) for the improvement of relations between the two countries.[18] This was not to be Iraq's last word at this stage on the possibility of a break in the hostilities. Early in October, after having blocked a UN attempt to halt the fighting, Iraq called for a cease-fire. This was intended to run from 5 to 8 October, but whatever the purpose underlying the move, the respite for those at the battlefront was brief. The Iraqi cease-fire ended after barely an hour.[19] It was supposed to begin at 5 a.m. on 5 October, on condition that the Iranians did not continue the war. President Bani Sadr of Iran said that his country was unwilling to respond, with the inevitable consequence that Iraq simply went ahead unilaterally and then cut the experiment short.

Iraq's attempt to give the Tehran government a quick bloody nose was only a partial success. Territory was gained, to be sure: Qasr-e-Shirin quickly, and a strip running down to Mehran. Khorramshahr, grimly dubbed Khuninshahr (city of blood) by some Iranians, was the subject of a good deal of contradictory reporting before it eventually fell[20] late in October. Despite a fierce battle,[21] Iraq failed to gain possession of the crucial oil centre at Ahwaz and cut off the supply of crude oil to the north. Late in September, it claimed to have taken Ahwaz,[22] but this was denied by Tehran radio. Despite some degree of battlefield success at Dezful—in the course of which Iraq captured a number of British-made Chieftain tanks which it put on display before foreign journalists at al-Amarah in the central sector[23] just as Iraq's own armour, mainly Russian in origin, was gathering for further fighting at Ahwaz—Dezful did not fall either[24] and the Iranians were far from having been routed. Where had the Iraqis gone wrong in Khuzestan? Some remarks made at the time by the defence correspondent of a major London weekly[25] are representative of a good deal of professional opinion expressed both then and since. Iraq's tactics, which bear the stamp of Soviet military training, were based on long-range shelling to soften up defensive positions in and around key towns. The classical sequel, which the Iraqis appear to have omitted, is then to send in infantry to mop up any remaining opposition.

There has been much conjecture about a lack of motivation at lower levels in the Iraqi formations, as contrasted with that displayed by the Iranian Revolutionary Guards, or Pasdaran, whose commitment and capacity to hold out against Iraqi superiority in equipment were important features of the war's early years. One factor with which the Baghdad planners had failed to reckon seems to have been the contribution to an enemy's performance of simple patriotism, the force which impels people to fight to defend their native soil against an invader, whatever they may have to say about the way things are run at home. This factor seems to have made a solid contribution towards forging the Iranian forces into a much more effective fighting machine than might have been expected in 1979. (It should be noted, however, that the Iraqis also, in later stages of the war when Iran was on the attack, were to demonstrate this characteristic in their turn.) Another explanation, also of a psychological nature, was faith—in God, in Islam, in their leader—bringing with it a preparedness to confront death in battle. This was widely thought to be the factor which attracted Iranians to the Pasdaran and others, young and old, to the *baseej*, or volunteer corps.[26] These rather unusual troops figured prominently in later engagements, sometimes being hurled against the Iraqis in what were widely described as 'human waves', recalling reports of the technique employed by the Chinese in the Korean war, in a bid to sweep away the Iraqis by sheer fervour and weight of numbers.[27]

Shortly after the invasion began, Iraq appeared to be making significant progress, but then unaccountably relaxed its grip. Not only did it fail to achieve all the gains on which it seemed to be calculating in embarking on the assault, but it sustained one particularly debilitating blow in the course of it. Soon after the Iraqi onslaught, reports were received that Iraq's two southern oil-export terminals close to Fao at the mouth of the Shatt al-Arab, Mina al-Bakr and Khor al-Amayyah, were under attack by Iranian warships.[28] The two terminals and the Fao tank farm were put out of action, depriving Iraq of the normal means of exporting the output of its southern oilfields close to Basrah. Fortunately it still had its fields in the north, with their pipeline systems leading to Turkey and Syria. These were to prove vulnerable as time passed: in Turkey to sabotage and in Syria to government action. In the meantime the scene was set for Iran to mount a major counter-offensive.

2
Five Thousand Years of Conflict

The outbreak of war in September 1980 between Iran and its principal western neighbour was not altogether without precedent. Apart from the battle of Qadesiyah[1] in AD 637 during the original advance of Islam, so frequently invoked by Iraq[2] as a source of inspiration in the more recent conflict with Iran, Persians and the inhabitants of what is today Iraq have been making forays into one another's territory for centuries.

During the first three millennia BC, two of the main concentrations of political power of the day, the Assyrian empire and Babylonia, lived in a state of near-permanent warfare, first with one another and then with the Persian kingdom of Elam, which at different times captured both Babylon and Ur and was itself in about 1100 BC laid waste by the Assyrians.[3]

The history of conflict between the peoples of this region over the centuries was interpenetrated by episodes in which it was subject to the imperial development of Rome, Macedon, Byzantium and the various peoples that streamed out of Central Asia into Persia and beyond. The best known of these were the Ottoman Turks.

The region's legacy from these centuries was not confined to models of conflict, however. In particular, there was the emergence of writing through the distinctive script called cuneiform, developed in Mesopotamia during the first three millennia BC, all within the framework of a civilization which also generated important advances in pottery and metal-working.

The build-up to Qadesiyah itself was long and slow. After a period of subjection to Assyrian and later neo-Babylonian rulers, Persia saw the emergence in the first millennium BC of the Achaemenid dynasty. Despite the defeats sustained in Greece,[4] which we know from Herodotus, the Achaemenids served as a major source of inspiration to the late Shah Muhammad Reza[5] in the formulation of both policy and propaganda. Then while the Sassanian dynasty was emerging in its turn, following the Parthians[6] who had swept away the Seleucid regime that ruled both Mesopotamia and the lands farther east after the break-up of Alexander's vast empire in the fourth century BC, the scene was set for the emergence of Islam.

This brief historical overview helps explain a number of later relationships. In AD 622 an inhabitant of the west Arabian town of Mecca (Makkah), Muhammad ibn Abdallah,[7] later to be hailed as the Prophet and exalted by visions he had experienced, embarked upon his celebrated flight from Mecca to Medina (Madinah), some 350 km away to the north. On his death ten years later, he was succeeded as leader of the community by one of his companions, Abu Bakr, who later became seen as the first of a line of caliphs (*khulafa* in Arabic). It was when the community of Muslims had progressed to the second caliph—another companion of Muhammad's named Omar ibn al-Khattab—that Omar's Arab followers, already having to confront the resistance of the Byzantine emperor to their efforts to expand their area of influence northward into Palestine and Syria, found their drive for conquest opposed by the Persian empire. This was now centred on Madain (or Ctesiphon), not far from the site of future Baghdad. At their first major encounter, Emperor Yazdegird came off the better. But in a second battle, at Qadesiyah, near Kufah in central Mesopotamia, a large force of Persians was solidly defeated[8] by a smaller body of Arabs, led by Saad ibn Abi Waqqas.[9] Saad later went on to take Madain and bring the Sassanian empire to an end, initiating in the process the substitution of Islam for Zoroastrianism as Persia's state religion. Even so, pockets of the earlier faith have survived into the present century.

In the meantime, the Muslim community was subject to complex stresses which were to prove significant both for the events of the immediately succeeding centuries in Mesopotamia and Persia and for those of the 1980s. The issue on which these tensions mainly centred was the succession to the caliphate. An important aspect of western Arabian society was the division of the Prophet's own tribe, the Quraish, into two main rival groups. These were the Bani Umayyah, whose head, Abu Sofian, was a particularly determined opponent of Muhammad during the early years of the propagation of his message, and the Bani Hashem, of which Muhammad was himself a member.

The second caliph, Omar, was assassinated in 644 after a reign which had seen, in addition to Qadesiyah and dramatic territorial expansion at the expense of Byzantium, an extremely important development within Islam. Foreseeing the need to elect his successor when the time came, Omar designated a body of electors charged with the task. In the event, the choice lay between a member of the Bani Umayyah, Othman ibn Affan, and Muhammad's first cousin, Ali ibn Abi Taleb of the Bani Hashem. The victor was Othman. What brought Othman's reign to a close was the outbreak of mutiny in garrison towns in Egypt and Mesopotamia, culminating in a march by mutineers on Othman's capital at Medina, where some of them later murdered him. Shortly afterwards, his defeated opponent Ali reappeared and was in turn acclaimed caliph. He was not without opponents, however. Some time before, a kinsman of Othman named Muawiyah, son of Abu Sofian, had been made governor of Damascus. A dispute quickly developed between Ali and Muawiyah, later erupting into open war, with Ali's newly chosen capital, Kufah, the focus of contention against Damascus.

Efforts by some of the faithful to arbitrate a settlement between the two men failed and Ali died in obscure circumstances in 661. Once again the issue of succession arose. On the Bani Hashem side, a prospective candidate was Ali's first son, Hassan, who appears to have despaired of achieving anything and was comfortably defeated by Muawiyah. The way was now clear for the emergence of the Umayyad caliphate, with Damascus as its capital. Under the new regime, succession was a straightforward affair. Muawiyah dealt with it by quite simply nominating his son Yazid in advance.

Next came what was to prove a particularly important development for the future evolution of Islam. The Umayyads' governor in Mesopotamia somewhat clumsily managed to provoke a revolt in the course of which Ali's second son Husain, a grandson of the Prophet,[10] and his complete entourage including small children were massacred at Karbala, not far from Kufah. This act of slaughter in 680 led to the polarization of Islam between, on the one hand, the supporters of Ali and his descendants (through whom they asserted the legitimate succession to the Prophet to run) and, on the other hand, the Umayyad caliphate establishment. The former took the name of Shi'a (which conveys the concept of a party of followers—in this case of Ali, whom the Shiites especially venerate, and by extension also of Husain) and the latter that of Sunnis (connoting the *sunnah*, the tradition relating the life-story and precepts of Muhammad, compiled from the Qur'an and the commentaries of the Prophet's contemporaries.[11]

The martyrdom of Husain, as the Karbala episode is usually described, is commemorated early in each Muslim New Year for the first ten days of the month of Muharram: in Shiite communities men form processions, bare to the waist, beating themselves with whips and chains until their bodies run with blood, chanting a recitation of the events of 680. Although it is not known for certain whether Yazid was present at Karbala, he is cast by the Shiites as the author of the incident, and the name is to this day a favourite Shiite epithet for a Sunni adversary—it was used in this sense during the 1980s war by the Tehran leadership when referring to that in Baghdad.

The distinction between Sunni and Shiite persists to this day. Well-defined communities of both sects are found in many Middle Eastern countries, sometimes living in rivalry, sometimes—as once in Lebanon—calmly coexisting; but always decidedly different.

Persia meanwhile was digesting the experience of conquest by the Arabs which had begun with Qadesiyah. One of its features was the widespread adoption of Arabic for certain purposes in Persian life[12] until the resurgence of Persian in the eleventh century AD.

By 750 the time had come for the caliphate itself to be transformed and for the Umayyad regime to be dislodged. The regime had been under strain for some time, not only up in the Syrian heartland but also in far-away Khorasan in northeastern Persia. This change, which was not provoked by any external force, was inspired by Muhammad ibn Ali, great-grandson of Abbas, an uncle of the Prophet, whence the name of the Abbasid dynasty which now succeeded the Umayyads. It is not perhaps surprising that the source of substitution should have come from the traditional opposition group, the Bani Hashem.

Muhammad ibn Ali promoted a campaign which used the medium of itinerant missionaries and was calculated to appeal to most of the people who for different reasons were discontented with the Umayyads. These included the Shiites, supporters of Ali, to whom Muhammad's campaign was particularly slanted and with whom it proved especially successful in both Persia and Mesopotamia. (This is not to imply that Persia was then, as now, populated mainly by Shiites. In the centuries following the Arab conquest, the dominant sect was of course Sunni.) Khorasan—and areas farther afield—were soon in open revolt, with radical figures like Abu Muslim[13] prominent.

The Umayyads were finally overthrown and the first Abbasid, Abul-Abbas, took the throne. It was the second of the new Abbasid caliphs, Abu Jaafar Mansur (754–75), who founded the city of Baghdad, a resplendent centre-piece for the empire until its sack by the Mongols in 1258 under Hulagu Khan. After this, a fugitive Abbasid prince was established in Cairo as a sort of emigré caliph under the protection of the Mamluk rulers there.

One of the first peoples to burst out of Central Asia and into the lands south and west of the Caspian were the Seljuks, a people of Turkish stock who subdued a broad band of territory from Afghanistan across to Syria, occupying Damascus and Jerusalem. Their path westwards can still be traced by the chain of distinctive buildings they erected on their way across.[14] They coexisted with the now somewhat depleted Byzantine empire but soon found themselves embroiled in the Crusader invasion at the end of the eleventh century, in the course of which were to emerge Baibars the Mamluk and the celebrated Kurdish warrior known as Salahuddin al-Ayyubi or Saladin.

By the beginning of the thirteenth century, Genghiz Khan and his Mongols were entering the picture. He created a massive Mongol federation in *c.* 1206, which in *c.* 1220 defeated the Caspian state of Khawarizm, then taking Samarkand and Bukhara. By the time of his death in 1227, Genghiz ruled over an empire that stretched from Persia across to China. His successors defeated the Seljuks in 1243, overcoming the Ismaili Assassins[15] at Alamut in 1256, and then swept away the Abbasids two years later. Finally they took on one opponent too many, receiving a resounding defeat at the hands of the Mamluks at Ain Jallud in 1260.

After this episode, the Mongol empire, of which a colourful picture is to be found in Marco Polo's account of his travels,[16] broke up into smaller political units, or khanates. In Persia, it was followed by the Il-khan dynasty. Although it contrived to remain fairly independent of the Great Khan far away in China, Persia had not seen the last of Mongol invasions.

Next to appear was Timur (or Tamerlane) in the 1380s, author of spectacular massacres and grandfather of the notable early astronomer Ulug Beg. The Timurid dynasty which followed gave way early in the sixteenth century to the Safavids, whose principal figure was that of Shah Abbas (1587–1629) and principal memorial the city that bears his stamp—Isfahan.[17]

The Safavids, claiming descent from Musa al-Kazem, the seventh Imam in the sequence of twelve in succession to Ali,[18] were fervent Shiites and gave the Ithna Ashariyah variant of the Shiite rite (highlighting the role of Imam Musa) something of the status of a state religion.

By now, the Othmanli (or Ottoman) Turks, another migrant group from Central Asia, were firmly established in Anatolia in succession to the Seljuks, having wrested Constantinople from the Byzantines in 1453. They soon assumed the role of Sunni champions against Shiite Persia, some of the time taking and retaking territory from one another. There is a tale that Sultan Selim, in setting off to do battle with his neighbours, once arranged to be equipped with *fatwas* (formal religious rulings) from senior Sunni *ulema* (divines of great learning in theology and Islamic law) proclaiming the slaying of Shah Ismail, heretic and unbeliever, to be a religious duty.[19]

In the sixteenth century the Ottoman regime made one important further step, defeating the Mamluks in 1516, which left the way clear for Ottoman domination of the main Arab-inhabited areas—Syria, Palestine, Egypt and the Arabian peninsula. (Suleiman the Magnificent was to take Baghdad in the course of the 1530s.) Finally, by a process which is not entirely clear, the Abbasid caliph al-Mutawakkil is said to have transferred to Sultan Selim and his heirs all rights to the caliphate.

After such a long and turbulent history involving Persia and its western neighbours, it is not surprising that the question of Iran's frontiers with the previous territories of the Ottoman empire has remained a live issue. Dr Khaled al-Izzi, in a sturdy defence of the Iraqi position on the Shatt al-Arab dispute published in 1981,[20] cites episodes of occupation of Ottoman territory in what is today Iraq in the sixteenth century and the signature of a peace treaty at Zohab in 1639.[21]

Another commentator on the troubles of earlier centuries was Lord Curzon, later to become Viceroy of India and British Foreign Secretary. Following a journey to Persia in 1889–90 of which he published an account[22] in 1892, he tells of the constitution in the 1840s of a Turco-

Persian commission pursuant to the Second Treaty of Erzerum,[23] concluded at the prompting of Britain and Russia:

> to prevent an impending collision between Persia and Turkey on the Kurdish border, Great Britain and Russia secured the appointment of a Turco-Persian Commission, to which were attached British and Russian representatives[24] and which met at Erzerum in 1843, in order to discuss and settle the frontier question. As a consequence of these deliberations, the Treaty of Erzerum was signed in June 1847.

As Lord Curzon points out, this was the second treaty to be signed at Erzerum. The first dated back to 1823.[25] Describing the approach adopted over the second, he says:

> [the commission] did not . . . attempt to demarcate a boundary but indicated a border strip some twenty to forty miles [30–60 km] in width, somewhere inside which the frontier line was understood to lie, the two governments being left to settle the question by agreement or force as they pleased.
>
> This somewhat uncourageous solution was confirmed in 1869 by a convention between Turkey and Persia in which each side undertook to respect the *status quo* until a settlement was arrived at.
>
> On many occasions . . . 1870 . . . 1873 [and] 1874 . . . disputes arose and armed collision was narrowly avoided.
>
> When in 1878 the European Congress met at Berlin to rearrange the map of Europe and generally carve up Ottoman territory, the question of the Turco-Persian frontier was again raised.

In fact, only a small change resulted. Another British commentator of the period, Lord Stratford de Redcliffe, in his authoritative treatment of nineteenth-century Ottoman affairs, *The Eastern Question*, gives a map illustrating the territories with which the congress intended to deal,[26] including the frontier between the two.

The fact that it has been necessary so many times over the years to attempt interventions to halt conflict and define stretches of frontier invites an explanation. There is an old Middle Eastern story about a frog (in some versions it is a fish) and a scorpion who meet on the bank of a river that both need to cross. In 1980 we ought perhaps to have asked ourselves whether this river might have been the Shatt al-Arab itself, while one of the characters was Iranian and the other Iraqi.[27] Although the frog is fearful of the scorpion's sting, they finally agree that the frog will swim across with the scorpion sitting on its back. 'Why be frightened?' exclaims the scorpion. 'Why should I do you any harm when it is you who are helping me?' The frog has no answer, so off they set. Needless to say, half-way across, the scorpion does indeed sting the frog and the pair of them perish. End of story, which is told and re-told, with the nationality of the scorpion, revealed in the punch-line, varying according to inclination and circumstance. This tale of perverse failure to act in terms of transparent self-interest seems a not unreasonable commentary on the predicament of two neighbours with a number of common concerns. Looking back to the period between the two world wars and even to that of 1939–45, any newcomer to the scene would have foreseen nothing other than a future of amicable and mutually beneficial co-operation.

Both Iran and Iraq were Muslim countries, both were ripe for economic development, both had significant oil resources being exploited by foreign concessionaires[28] (they were at one stage to co-operate over the development of a small oilfield straddling the frontier near Kermanshah) and both had restive Kurdish minorities. In addition, they shared a concern for good relations with neighbours—or so some might infer from their both being signatories to the 1937 Saadabad Pact[29] in company with Turkey and Afghanistan. Moreover, moving from the official to the private level, there was a significant and regular cross-border movement of ordinary individuals, if only because of the existence in both countries of Shiite Holy Places—in particular, Qom and Mashhad in Iran, Najaf and Karbala in Iraq—which attracted visitors, whether as a form of pilgrimage or to bury their dead.

The expectation of good relations was not to be fulfilled. Nor at any time after the Second World War did they seem likely, despite several important developments relating to areas of common interest. First, there was the 1960 agreement, to which both Iran and Iraq were parties, whereby several major oil-producing states set up OPEC. Second was the frontier agreement between the two countries signed at the 1975 OPEC summit in Algiers. This came at a time when several years of large oil-revenue increases had created an exceptionally propitious atmosphere for reaching some kind of understanding. Relations nevertheless remained strained.

The 1975 Algiers agreement had been necessitated by several factors: Iraqi charges of Iranian moral and material support for Kurdish insurgents; widespread Arab resentment (which Iraq sought to champion) of Iran's seizure of the three islands down the Gulf with apparent British collusion in 1971; friction over the Shatt al-Arab concerning both frontier demarcation and responsibility for river traffic; and Iraq's 1972 treaty of friendship with the USSR, a neighbour regarded by Iran with some disquiet.

The Algiers agreement failed to work out to the satisfaction of either side. Along the inland, mostly mountainous boundary, despite the efforts of the joint delimitation commission, much remained to be clarified and was still causing difficulties in 1980.

In the Shatt al-Arab, Iran had been accommodated in 1975 by Iraq's acceptance of the *thalweg* principle[30] as applicable throughout the length of the waterway (and not in part only, as under the provisions of Constantinople (see below) in 1913 and the treaty negotiated between the two countries in 1937).

The frontiers of Persia have always been somewhat imprecisely defined. Back in the thirteenth century, Marco Polo[31] described Bukhara as the finest city in all Persia and Tabriz as the most splendid city in the province (of Iraq), neither of which corresponds to our contemporary understanding. The following century, Ibn Battuta[32] related how on the two sides of the southerly part of the present-day frontier were two distinct tracts of 'Iraq': Iraq al-Arabi, which was akin to today's Iraq; and Iraq al-Ajami (or Persian Iraq), inhabited by Persian-speakers and including the city of Isfahan.

In addition to the Erzerum and earlier treaties, there was a long history of tension in the region. In 1911, for example, the Tehran Protocol[33] was concluded following a Persian appeal to the mediating powers, Britain and Russia, over Turkey's alleged violation of the 1847 Erzerum Treaty. The protocol provided for a delimitation commission to be formed of representatives of the two governments. This started work the following year. Nevertheless Turkey and Persia were swiftly back in disagreement. The Constantinople Protocol of 1913[34] was an attempt by Britain and Russia to overcome the problem that had developed in 1911. The agreement defined the frontier and was signed by all four states. Next came the Iran-Iraq treaty of 1937. In the main a confirmation of the content of earlier agreements, it secured Iran an anchorage in the Shatt al-Arab opposite the oil-loading berths at the Abadan refinery.[35] This was followed by the 1975 Algiers agreement,[36] which Iraq abrogated in 1980.[37] A further series of frontier incidents occurred during the course of 1980.[38] These were fairly continuously reported over the months preceding Iraq's invasion in September by the BBC monitoring service and by newspapers like *The Times*, which on 20 September, just three days before Iraq moved in, reported an Iranian attack with tanks and jets.

Incidents of this kind were not the only contribution to tension over this period. There were many other examples:

(a) an attempt to assassinate future Iraqi Foreign Minister Tareq Aziz[39] in Baghdad;
(b) Iraq's expulsion of some thousands of Iranians resident in Iraq;[40]
(c) reports that Iraq was sending arms to the Arabic-speaking minority in Khuzestan;[41]
(d) manifest attempts by Iran to intervene in Iraqi domestic affairs. The BBC monitor report of 18 June 1980 carried a transcript from Tehran radio of two days earlier, containing the declaration: 'We hope that . . . [Iraqis] will follow the orders of the Imam and overthrow the ill-fated Baathist government which is controlled by a small clique';
(e) the sentencing to death of a number of people in Iraq on charges of spying for Iran and Israel;[42]
(f) certain formal steps like Iran's call-up of reservists and earlier the recall of its ambassador from Baghdad and placing its army on alert to a degree that the Iranian air force commander at the time described as 'unprecedented';[43] and
(g) the execution in April 1980 of Imam Muhammad Baqer Sadr, the prominent Shiite cleric, whose house arrest the previous year had aroused Ayatollah Khomeini's personal concern. Pressure on the Shiite community was to continue, despite well-publicized conciliatory gestures. In 1983 several clerics (variously reported as five and six) of the al-Hakim family were executed and their relations imprisoned.[44]

Another factor was widely seen as being Iraq's concern to extend its influence in Arab world affairs generally, an aim not unlikely to be served by administering a lesson to Iran.

3
The Adversaries

The present chapter turns to the protagonists themselves as they appeared to the outside observer in 1980, when continuous fighting began. It will examine both their societies and the resources (human, economic and military) with which they embarked on the war.

Iran: Basic Resources

Iran was much the more populous of the two countries, with 33.6 million recorded in 1981 against Iraq's 12 million.[1] It was just in the process of digesting the Islamic revolution which had set the last of the Shahs, Muhammad Reza, on his unhappy way to exile, illness and death, the last finally overtaking him in Egypt.

Modern Iran was largely the achievement of efforts of will on the part of two Shahs—Muhammad himself and his father Reza. Reza Shah was an ex-Cossack officer who in 1925 seized the throne from the last of the preceding Qajar dynasty. With his forceful handling of the clergy and zeal for modernism, he bequeathed an example which the luckless Muhammad was obliged to try to emulate. And try he did. For all the criticism and resentment he aroused, he put in a formidable effort to make of Iran a state its neighbours would envy and the outside world respect. To this end, he boosted the standing and weaponry of the armed forces on whose support his prestige as a monarch substantially rested. He fought for higher oil revenues to finance this and the development programmes that would transform Iran's economy. To complement this transformation he tinkered with the reform of society and institutions. His 'White Revolution' included the distribution to peasants of title-deeds for the land they worked; a campaign to eradicate illiteracy; and some experimentation with parliamentary-level political dialogue—at one stage an instant opposition party,[2] later a single-party[3] attempt to institutionalize his dream of nationwide support with distinct and opposed wings of opinion.

In the process of attempting all this, the Shah inevitably attracted criticism. It came from many quarters, both domestic and foreign, for his megalomaniac, as it was often characterized, insistence on Iran's 'imperial vocation' of which in 1972 he celebrated the 2,500th birthday.[4] It also came from the clergy, who disliked the White Revolution generally,[5] and from the liberals whose thunder he had to some extent stolen. That the White Revolution was massively acclaimed by referendum in 1973[6] did not commend it any the more to the clergy. Further criticism came from foreign commentators who found his espousal of Western ideas too selective by far, taking him to task for his autocratic style, for underpinning his regime with the repressive Savak security apparatus, his sybaritic leisure-image—notably featuring snow and sun at St Moritz—and his failure to uproot corruption in his entourage. The severest revilement, however, probably came from the mob in the street[7] and the clergy.[8] Others who may well have had reservations about him (though we shall probably have to wait some time before their reflections can be made available to us) were the oil-company spokesmen to whom he advanced new facts of life[9] during the earnestly disputed discussions between OPEC and the major companies in Tehran in December 1973.

Late in his reign, the Shah was even said to have been impressed by the liberalizing efforts of King Juan Carlos in Spain. If these reports are correct, the idea led him nowhere; in the end, weakened by the effects of his illness and, some have contended, with his personality affected by attempts to treat it, he lost his grip and abandoned the struggle.

The revolution that toppled this edifice in 1979 was not the work of any single organized movement. Opposition to the Shah's regime was disparate—fragments of the old National Front, a survival from the times of Mossadeq in the 1940s and 1950s; a miscellany of resurgent Islamic and minor left-wing groups; the pro-Moscow Tudeh Party; disappointed businessmen who perceived the hand of the Shah's family behind every deal they failed to clinch; and smaller traders who resented the Shah's busying himself with high prices in the bazaar. Disparate and haphazard as it was, one thing is certain: popular discontent with the system had been general for several years. In the course of return visits to Tehran and Isfahan in 1977, I was continually struck by an air of impending serious troubles like those I remembered engulfing the country in 1963.[10] Many well-to-do Iranians, convinced of imminent disaster, were sending their liquid assets abroad, some actually accompanying them in person.

Part of the resistance took the form of urban terrorism with the murder of foreigners in Iran and an attempt in 1977 in France on the life of the Shah's sister, Princess Ashraf.[11] Other important elements included the student body at Tehran University and oil-industry workers in Khuzestan (where most of the oil operations were located), who were intermittently on strike for much of 1978.[12] Just before this, there had indeed been something of a re-enactment of 1963, with demonstrations in the holy city of Qom ending in shooting; processions elsewhere in sympathy with those killed; and the imposition of martial law. The last of these was one of the principal causes of the oilfield strike, whose most important effect was a dramatic fall in crude oil output, from 2 billion barrels (bbl) in 1977 to 0.5 billion in 1980.[13] A not insignificant push was given to this development with the decline in morale among expatriate workers in the oil industry following the assassination of a senior oil-company secondee near Ahwaz late in 1978.[14] This fall in oil production jolted the process of economic management, particularly since the leadership had become accustomed to a high level of revenues after the OPEC price increases, first during the campaign to this end in the early 1970s and then as a by-product of the 1973 Arab-Israeli war.

The effects of these increases were not always positive for the oil-producing countries. Although they conferred a degree of freedom from financial constraints in decision-making, they had the effect of creating areas of over-heating in the economy. Price inflation plagued the distribution of imported goods and was aggravated by bottlenecks in ports like Khorramshahr and beyond. By 1984 the *Financial Times* issue of 13 February was reporting Iran as cutting imports since the country's ports and internal distribution services could no longer handle the flood of goods destined for the country. At Bandar Abbas, for example, 90 ships were queueing to discharge.

The post of prime minister under the Shah was a precarious one. Sharif-Emami took on the job twice, the second time in 1978; Mansur was assassinated in 1965, like Razmara in 1951; Amir Abbas Hoveyda,[15] who replaced Mansur, had a long run but was eased out in 1977 to give way to Amuzegar,[16] a successful OPEC hawk as previous oil minister. In the end, Hoveyda was tried by a revolutionary court and executed,[17] but the story of the succession of prime ministers continued, with disturbances breaking out all over the country, up on the Caspian as well as in the major centres. After martial law had failed to bring them under control, there came a last military experiment with the designation of General Azhari to succeed Sharif-Emami. This move proved a mere stopgap until the appearance of the last pre-republican premier, the moderate conservative Shahpur Bakhtiar.

Meanwhile at Neauphle-le-Château, near Paris, a figure sat, biding his time. Ayatollah Ruhollah Khomeini had spent much of his life combining a climb towards the summit of the Shiite hierarchy with the elaboration of a coherent set of ideas about what was wrong with both contemporary Iran and much of the outside world. The three principal figures in his demonology[18] proved to be the Shah, Israel and the United States, and in a sermon against this trio in March 1963 he had done much to trigger the disturbances of that summer and the process that led to his banishment to Najaf in Iraq.

In Najaf, Khomeini had begun a propaganda campaign against the Shah and his regime by circulating audio cassettes[19] containing his pronouncements. He had left Najaf in 1977, finding asylum in France and a new operational base from which to beam his messages to the populace, such as calls for the creation of the revolutionary 'komitehs' that sprang up 'in many parts of the capital'[20] and appeals to the army to desert.

By the end of 1978 renewed demonstrations in Isfahan and Tehran (and elsewhere) had demoralized both the Shah's government and the Americans, who were busy trying to salvage something from a situation that was clearly going badly wrong. In mid-January, a disconsolate Shah gave up and left.[21] Rioting continued, however, and immense crowds made their way to Mehrabad airport for Khomeini's return from France on 2 February.[22] Descriptions of the occasion are reminiscent of Lenin's legendary arrival at the Finland Station in 1917.[23] The next two years were no less eventful than those preceding them. In March 1979 a republic was formally proclaimed and a moderate National Front figure, Mehdi Bazargan, designated provisional prime minister.

The apparatus that emerged from the period of constitutional ferment that now ensued featured an elected national assembly (Majles) as before; a Revolutionary Council composed of senior clerics and some lay figures involved in the Shah's overthrow; a president of the republic; and a Council of Guardians, whose task was to pronounce upon the Islamic qualities of any candidate for office and on any proposed course of action. Finally, there was a device to institutionalize Khomeini—informally known as the 'Imam', the community's leader in prayer—in a role of supreme jurisconsult/general arbiter, Velayat-e-Feqih. At about this time there also emerged the Islamic Revolutionary Party (IRP), to provide a banner under which approved candidates could run for election to the Majles, and the Revolutionary Guards or Pasdaran-e-Enqelab-e-Eslami.

There were still problems to overcome. Guerrilla bands overran the principal military barracks in the Tehran area and the arms it contained appear to have ended up in hands of varying political allegiance. The army and air force leaderships were at odds with one another, and both morale and efficiency were low following the desertions prompted by Khomeini and the revolutionary purges of the officer corps. There were also queries over the battle-worthiness of aircraft and other major military hardware due to maintenance backlogs and shortages of spares. Finally, there remained differences of approach between the various forces which had contributed to the Shah's downfall. The Tudeh Party appears initially to have decided to try to work within the new system, but was later subjected to pressure, especially in 1984.

The new regime's most determined opponents eventually emerged as the Mujahedin-e-Khalq, led by Mas'ud Rajavi who established a base in Paris for some years and was reported in 1986 to have made his way to Baghdad.[24] In addition, there were differences within the circle of senior clerics. Khomeini found Ayatollah Beheshti overly ambitious[25] and took steps to block his path to the presidency. The Imam was also reported to have encountered disagreement from other ayatollahs—from Taleghani[26] and Shariat-Madari[27] (both of whom have since died), the latter being accused of involvement in an attempted coup. As if this array of problems were not enough, the regime took a new one upon itself later in 1979, when the US embassy on Takht-e-Jamshid Avenue was stormed by students and its occupants taken hostage. Both parties, Iran and the United States, were damaged by this episode. Iran's oil sales to the US and assets there were frozen by the Carter administration, whose later attempt to mount a Hollywood-style rescue ended in the desert at Tabas[28] in an ignominious collision of aircraft. There was also pressure for some degree of autonomy from minorities in Kurdestan, Azerbaijan and Baluchistan, from Turcomans in Khorasan and from Arabic-speakers in Khuzestan. (This last was ignored by Iran, which perceived it—perhaps with some accuracy—as having been staged by Baghdad from the outset.)

On the positive side, oil production, after its 68-day export shutdown, was functioning more or less normally and managed to register a full-year average by the end of 1979 of 3.1 million barrels per day (b/d).[29]

Iraq: Basic Resources

In the case of Iraq (area 434,924 sq. km against Iran's 1,648,000),[30] its society and economy had spent a number of years digesting the effects of coup and revolution,[31] together with a good deal of supplementary drama. This meant that, by 1980, the country had developed into a modern territorial state under strong central control.

Although Iraq had achieved formal independence in 1932, with the exit of the British administration installed under the League of Nations mandate, there remained for a number of years a substantial British presence. This included RAF stations at Habbaniyah and Shuaibah; a British company developing the oil resources in Iraq's subsoil; British archaeologists[32] investigating Iraq's ancient past; and a solid nexus of close personal ties between Britain and the Hashemite regime.

This being so, and given the post-1918 spread of nationalist sentiment in much of the Arab world, it is not surprising that the nervous and physical energy of many Iraqis was channelled into the shrugging-off of this association. One such attempt came to nothing during the Second World War, when a military coup mounted under the pro-Nazi Rashid Ali al-Gailani was defeated.[33] Nevertheless the underlying sentiment had a good deal to do with Iraq's inability to stand by the treaty it had signed with Britain at Portsmouth in 1948, and resentment aroused by post-war developments in Palestine did little to help. Britain abandoned the Habbaniyah base in 1959 after Iraq left the Baghdad Pact in the wake of the previous summer's revolution. Iraq's nationalization of its oil industry was to be a step-by-step process, but the critical step had to wait until 1972.

The first republican regime under Brigadier Qasem lasted only until 1963, when a joint coup by pan-Arab nationalists and members of the Arab Socialist Baath Party brought that combination to power. In 1968 another coup brought in a purely Baath regime in its place.[34]

During intervals between changes of regime, life in Iraq was often violent and precarious. In Mosul in 1959 there was a revolt against the new central government. The same year a series of dramatic trials was held in Baghdad of enemies of the revolution. In 1963 sporadic fighting broke out between elements of the regular army and the National Guard. For much of the period from the 1958 revolution to the outbreak of war with Iran in 1980, there was a well-nigh continuous insurgency in the Kurdish-inhabited areas of the north, a situation exacerbated by Iranian support. In 1968 arrests of suspected opponents of the regime were commonplace; the following year there were spectacular trials and executions of men accused of spying for Israel. Then in 1970 a coup attempt by an army/police combination was followed by more executions and imprisonments. Other sources of unease were the unfulfilled aspirations of the minority Shiite community, mostly concentrated in the region south of Baghdad, and the occcasional stirrings of the Communist Party which had gained strength during the Qasem period and become solidly represented in the army until efforts to purge them in the 1970s. In 1979 there were reports that a ministerial reshuffle had been accompanied by charges of a coup attempt fomented by a foreign power, culminating in a number of executions.[35] As Iraq moved through the 1970s the Baath regime became well settled in power, with the gradual emergence in key positions of a number of individuals from the town of Tikrit on the Tigris, north of Baghdad.[36] One such was General Ahmad Hassan al-Bakr, president of the republic from 1968 until 1979, when he was succeeded by his kinsman Saddam Hussein; others were Defence Minister Adnan Khairallah and Saddam's half-brother Barazan, head of the *Mukhabarat*, or Internal Intelligence.[37]

The Baath was no overnight creation, but went back to the 1940s when it was conceived by the Syrian duo, Salahuddin Bitar and Michel Aflaq, who later fostered branches in a number of Arab states. The two countries in which the movement has taken root are Iraq and Syria, each of which now claims leadership of the movement on the pan-Arab level and dismisses all similar pretensions on the part of the other. The consequence has been fierce rivalry between the two in all areas. This, in turn, is reflected in conflict in different contexts between groups linked with Iraq and Syria respectively, as in the civil war in Lebanon and within the Palestine Liberation Organization (PLO). It is sometimes said that in the light of the historical tension between Umayyad Damascus and Abbasid Baghdad, things could hardly have been expected to turn out otherwise.

There was, however, one issue on which both of them had similar views: President Sadat's peace-seeking journey to Jerusalem in November 1977 and its sequel at Camp David the following year. An Arab summit conference met in Baghdad to consider its impact, finally consigning Egypt to political quarantine. The outcome for Iraq and Syria was an effort at reconciliation, including the prospect of political union between the two, an elusive notion which collapsed in 1979 after many months of detailed and fruitless negotiation.[38]

In Iraq as in Iran, the country's backbone was the oil industry, now entirely under Iraqi control, though a number of foreign companies concluded service contracts with the Iraq National Oil Company (INOC). One of them was the Brazilian state oil concern Petrobras, which discovered significant accumulations in the south of the country, notably the Majnun field; another was the Soviet Union, with which Iraq concluded a treaty of friendship and co-operation in 1972.[39]

Material Resources

We shall now turn to the material resources—petroleum, financial assets, military hardware and its framework—available to the two belligerents at the outbreak of war. The fundamental element was, of course, oil. It emerged early on that each side was determined to deny the other the means of generating enough income not only to prosecute the war but also to govern the country. This appeared quickly as a cardinal element in Iran's strategy when Iranian naval craft, as we have already seen, bombarded the southern Iraqi terminals. As the war proceeded, Iraq attempted to do the same to Iran's terminals and bring Iranian exports to a halt.

Of the two oil-production systems, the more prolific was undoubtedly Iran's. By 1977 its installed production capacity was rated in industry circles at approximately 6.9 million b/d.[40] In 1974 real output was 6.02 million b/d,[41] a million or so up on 1972. The peak output level—6.678 million b/d[42]—was reached in late 1976.

Most of this capacity was in the Khuzestan area, where prospecting began in the early years of this century, culminating in the discovery in 1908 of the first of a cluster of oil-bearing structures identified in a northwest-southeast trend on a flank of the Zagros mountain range.[43] This delighted Mr Winston Churchill,[44] who as First Lord of the Admiralty was anxious to secure the Royal Navy's fuel supplies for the war which seemed imminent. Among the largest of these structures are the country's best-known fields, Agha Jari and Gach Saran.[45] Output was channelled in two directions: an amount varying from year to year but within the broad range of 500,000–1 million b/d went to the refineries at Abadan, Tehran, Isfahan and elsewhere inland; the balance went for export via the main terminal at Kharg Island. There were other Iranian terminals in the Gulf, at the islands of Lavan and Sirri, originally bases for offshore oil exploration and later adapted for export, rather like the islands of Halul and Das on the Arab side of the Gulf.

By 1980 Iran's total production capacity, weakened by reservoir depletion, natural pressure decline and a backlog of work on maintenance programmes accumulated over the preceding years,[46] was assessed at just 4 million b/d. The annual revenue generated by this apparatus over the years preceding the war had built up from $482 million in 1964 to $1,109 million in 1970. Then through the decade of price hikes, it registered $17.5 billion in 1974, $21.6 billion in 1977, but relaxed to $20.9 billion in 1978 and $20.8 billion in 1979, crumbling to $11.6 billion in 1980.[47]

In Iraq the situation was less comfortable, with proven reserves[48] of some 31 billion bbl against Iran's 58 billion bbl, though Iraq later in the war claimed higher levels.[49] Most of the production capacity had been inherited from the nationalized Iraq Petroleum Company (IPC) group of companies. Some of this capacity was in northern parts of the country, subject over the years to Kurdish insurgent activity; some in the south near Basrah. The biggest field was the oldest,[50] discovered in the 1920s at Kirkuk, north of Baghdad, where the foothills of the Zagros[51] range extend across into Iraq and geological conditions have certain common features with those in Khuzestan. Production from the northern fields had always been evacuated through pipelines to the Mediterranean. By 1980, with the IPC 'H' line to Haifa long since closed, there remained the 1.4 million b/d 'T' line to Tripoli on the Lebanese coast, with its branch off to the Syrian port at Banias. To supplement this there was now a 0.7 million b/d line through Turkey to Ceyhan. The output of the southern fields was exported through the terminals on the Gulf which were to be demolished by enemy action at the outbreak of war. To the fields brought in by IPC had been added others, discovered or developed by the national oil company INOC, under new-style arrangements, notably the Brazilian discovery Majnun,[52] prominent in the fighting near Basrah in the 1980s. Overall production in Iraq moved from 1.548 million b/d in 1970 to 3.3 million in 1979.[53] Annual revenues rose from $521 million in 1970 through $5.7 billion in 1974 to $20.3 billion in 1979.[54] For the first year or so of the war, oil and gas operations in the Gulf, leaving aside the damaged Iraqi terminals, continued very much as in the 1970s. Conscious of the uncertainty attending relations with its Middle East neighbours, however, Iraq had set up contingency arrangements. A 1 million b/d 'strategic' pipeline had been installed so as to enable the southern fields' production to be sent out via the northern pipeline system through Syria, or the reverse;[55] as it later proved, a sensible precaution.

The classic source for data on the two belligerents' respective financial positions, the IMF's *International Financial Statistics*, does not present a complete picture. Iraq appears—perhaps because of security considerations—to have withheld some of the data it would normally have reported, leaving gaps in the IMF's tabulations which have to be made good from less direct sources like the analyses conducted by the Bank for International Settlements (BIS) and economic commentators. What seems to emerge is that Iraq's foreign-exchange reserves, which amounted to $20 billion in 1980,[56] had reached $30 billion the following year.[57] Iran, more awkwardly placed, had reserves in 1980 of just $10,223 billion[58] and debts inherited from the Shah's time[59] of $6.5 billion.

In the area of defence, the foremost claimant for government funds in time of war, the picture available of the two belligerents' military machines, like that of their finances, is not always very clear. Although the subject received some coverage in the course of routine reporting on the war by journalists with access to the usual range of sources—specialist spokesmen in government information agencies, defence attachés in various embassies, the occasional interview on the field of battle—the main authorities are the specialized bodies which monitor developments in the strategic field, the international arms trade, and so on.[60] Although the figures do not always agree, a serviceable, if somewhat impressionistic, picture emerges.

In Iran the Shah, mindful of Iran's imperial past, had built up a formidable military machine. Although it was never in his time subjected to the trial of a major war, it was given one partial blooding. To preempt the danger of a revolutionary regime dominating the nearby Musandam peninsula,[61] the Shah in 1973 sent a contingent to help the Sultan of Oman put down an insurrection in Dhofar province prompted[62] by its communist neighbour, South Yemen. The principal elements in the 1980s line-up identified by the International Institute for Strategic Studies (IISS)[63] were some 12 divisions, 3 of them armoured, plus an airborne brigade. In addition to the total of 285,000 troops can be reckoned a certain number of Pasdaran, the Revolutionary Guards (later reported as numbering 250,000) and a similarly uncomputed number of *baseej*, the highly motivated volunteer formations combining adolescents with the over-40s.

Armoured fighting vehicles varied in origin: they included the US M-47 and M-60 and the British Chieftain. There was no such diversification of source in the Iranian air force. The main work-horses were US-supplied F-4 Phantoms and F-5 Tigers, while the stars of the line-up were the rather newer Grumman F-14 Tomcats. Later sources of aircraft and tanks included the People's Republic of China.

This shining array was not necessarily translated into flawless military effectiveness. Stories began to emerge about maintenance and logistical problems,[64] to the effect that only 20 of the original 84 Tomcats supplied were operational by 1981[65] (and no more than 40–50 aircraft altogether).[66] Spares were short, which drove Iran out into the underground arms market to make good the shortages, a process which ultimately led to the Irangate scandal in the US.[67] Well before that story reached the newspapers, there were reports 'according to well-informed European shipping sources'[68] of Israel sending US aircraft spares to Iran.

Iraq's armed forces were also solidly equipped, thanks to the atmosphere generated by the Iraq-Soviet treaty. The 1980 line-up of the regular army identified by IISS consisted of eight divisions, two of them armoured, and a number of independent brigades, together with some People's Army (al-Jaish al-Shaabi) formations of part-timers under the command of the regime's second most prominent figure, Taha Yasin Ramadhan, estimated in 1980 to number some 75,000. These forces were gradually grouped into the superior formations occasionally identified in government communiqués but more usually mentioned by external sources such as the BBC monitor reports.[69] Some seven army corps formations were identified in this way: the First in the north, the Second in the centre and the Third, Fourth, Sixth and Seventh in variously designated tracts of the southern sector of the front. It was never clear whether there was actually a Fifth or whether the whole array was simply devised in order to confuse Tehran.

The hardware was Russian, amounting to some 1,700 assorted tanks: the T-54, T-55 and T-62 at this stage, as it was only later in the war that Iraq received supplies of the newer T-72. Minor equipment and spares were to come, in addition to the USSR, from Egypt, Brazil, Chile[70] and some other suppliers including, controversially,[71] Greece. The pattern for aircraft was mostly the same. Apart from a few ageing British Hawker Hunters, the main workload was borne by the Tupolev T-22, the Ilyushin Il-28 and the MiG-21. It was not until well after the war began that Iraq started to diversify in the direction of France, with the Mirage and Super Etendard, both capable of delivering the much-discussed Exocet air/surface (AS) missile. Later, Iraq also received the MiG-29 Fulcrum.[72] By the time IISS brought out their 1984/85 issue of *Military Balance*, updating the position at the war's outbreak, Iraq was reported as having five Super Etendards. By 1987/88 the overall army head-count was 642,500, including 6 armoured divisions, 4 mountain divisions (to strengthen control of the north) and 15 People's Army brigades. There were also 15 Soviet SS-12 surface/surface (SS) missiles.

Gradually, as the war proceeded, both sides equipped themselves with a varied range of missiles, all with conventional high-explosive warheads.[73] While there is no evidence that either attempted to widen its options to the extent of shopping around for nuclear alternatives, both certainly took steps in the direction of nuclear-reactor installation, whether for research or power generation. Such a course is often open to the construction that nuclear weapons may be developed at some later stage, as at least a secondary objective. This appears to have been the conclusion reached by the Israeli government when it launched its air attack on Iraq's French-built research-scale Osiris reactor at Tuweithah near Ctesiphon in June 1981.[74] This story is continued in Chapter 13, where 1991 was to see a close UN interest being taken in Iraq's latter activities in this field.

Iran also had a nuclear plant under construction by a European firm, the West German Kraft Werke Union, at Bushehr on the Gulf coast.[75] This time it was the specialist defence press that placed the construction of military adaptation on the venture.[76] A separate report[77] indicated that Iran may have been working on the development of such a weapon on and off for a good decade. Tales of this sort continued to emerge; in early 1989, for example, *The Times* reported a secret programme[78] on the part of Iraq to build nuclear warheads for a strategic missile being developed simultaneously with technical assistance from Pakistan and finance from Saudi Arabia.

4
Interested Third Parties

By 1980 the succession of troubles along the Iran-Iraq border, combined with uncertainty as to the international behaviour to be expected of post-revolutionary Iran, was provoking a good deal of disquiet in the outside world, particularly among the Organization for Economic Co-operation and Development (OECD) states. They were in many cases consumers of Gulf oil and anxious on two counts: first, over the physical security of supplies, lest conflict in the region should provoke a breakdown in the outward flow of oil; and second, over price increases, which would bring in their train inflation, unemployment and general economic stagnation.[1] The Western world had already had a foretaste of this combination of factors following the price increases in the wake both of the 1973 Arab-Israeli war and of the Iran revolution itself in 1978–79. OECD and its related organization, the International Energy Agency (IEA), were still trying to digest the lessons and, in the case of IEA, devise mechanisms to avert a repetition.

One major OECD figure had already had momentous things to say about the Gulf. US President Carter, not destined to end his days with the happiest memories of the region, declared in 1980 that:

> Any attempt by an outside force to gain control of the Persian Gulf region will be regarded as an assault on the vital interests of the United States. It will be repelled by use of any means necessary, including military force.

This philosophy, which became known as the 'Carter doctrine',[2] persisted into the presidency of his successor and became part of the conditioning material at play in later Washington reflexes.[3]

In May 1984[4] President Reagan declared that:

> Neither we nor the Western world as such would stand by and see the straits of the Persian Gulf closed to international traffic.

In 1980 the US response to conflict at the head of the Gulf was twofold. First, a quartet of AWACS aerial surveillance craft[5] was dispatched to Riyadh. Second, the concept was aired of a 'rapid deployment force' (RDF), which might be brought into play in the Gulf as in any other theatre in which an emergency erupted outside the US.

AWACS aircraft, resembling ordinary Boeing 707s but each with a large metallic mushroom mounted above the fuselage, flew off from Riyadh's dual-purpose airport every morning, scanning both the Arabian peninsula and the war area to the east and north. The former part of their duties aroused a protest from Libya's Colonel Qaddafi during the pilgrimage to Mecca that year: unbelievers were traditionally forbidden from overflying the Holy Places in Saudi Arabia's Western Province, the Hedjaz. As to the deployment of a RDF, this would entail a mixed air and sea operation under the eye of the US Central Command based in Florida, to transport airborne, marine and other formations from the US.

As the war proceeded, Western uncertainty increased as to what developments to expect. The unease was heightened by reports of bellicose pronouncements or writings (as when the Tehran daily *Kayhan* printed an editorial advocating the aerial bombing of Arab ports on the Gulf serving as transit points for munitions destined for Iraq). The State Department reacted to this hint of interference with peaceful navigation by expressing the US administration's 'grave concern'.[6]

The following year, there was more grave concern, this time in Congress, following reports of an aerial battle over the Gulf[7] involving US-built aircraft, Iran's F-4 Phantoms and Saudi F-15 Eagles. This prompted fears that US personnel might be drawn directly into the politicians' nightmare—someone else's war, the spectre of another Viet Nam. This concern may to some extent have been justified: according to a newspaper report,[8] one battle in which an AWACS aircraft was involved, albeit peripherally, nearly became several.

Meanwhile 21 industrialized countries, gathered under the auspices of IEA, had elaborated an oil-sharing arrangement to be invoked in the event of any disruption which caused a shortfall of 7 per cent or more in a member country's normal supply arrangements. (The percentage figure was critical, based on past experience; in 1979–80, a forecast 5 per cent shortfall in supplies worldwide precipitated the price rise.[9] It was the Saudi oil minister, Shaikh Ahmad Zaki Yamani,[10] who pointed out that a cut-off of Gulf oil would raise prices and 'seriously affect the Western economic recovery'.

As far as the two belligerents were concerned, the US position was fairly clear-cut. Relations with Iraq had been broken off at the time of the 1967 Arab-Israeli war and US diplomatic business in Baghdad was handled by a US Interests section at the Belgian embassy[11] (headed by the author of an authoritative work on Kurdish affairs, William Eagleton Jnr). Relations were restored later in the war, following a visit to Washington by Deputy Premier Tareq Aziz in November 1984. In the case of Iran, relations with the US could hardly have been worse, in the light of the hostage-taking at the US embassy.

The USSR's position was also reasonably straightforward. As far as Iraq was concerned, it enjoyed a long-standing treaty of friendship and co-operation with the Soviet Union. The Baghdad government was duly appreciative of the fruits of economic co-operation, notably over oil-resource development, and a flow of military supplies. In the USSR, however, there were reports at the beginning of the war that the Soviets were unhappy that a war had broken out at all and that a friendly state, Iraq, was at war with a regime that had conducted a revolution of which the USSR basically approved—and using Soviet arms into the bargain. Moscow was said at this stage to have been somewhat sluggish over the provision of spares and replacement items. Evidently, the Soviets soon relented, as sightings of the advanced T-72 tank were noted soon after the war began,[12] and visits to the USSR by the Iraqi leadership began to be registered once again.[13]

Soviet relations with Tehran, apart from the general benevolence with which Moscow instinctively viewed the revolution, were more complex. Iran's new leadership tended not to accord Moscow any particular credit for its rivalry with the 'Great Satan', because of, first, the image of godlessness that Moscow enjoyed in Iran and, second, the presence of Soviet troops in neighbouring (and predominantly Persian-speaking) Afghanistan. The two countries had gone through an up-and-down relationship over natural gas exports from Iran to the USSR. The atmosphere was not improved when in 1984 Iran outlawed the pro-Moscow Tudeh Party and expelled a number of Soviet diplomats.[14] On the other hand, one source of satisfaction to the USSR, as one of the world's major gold producers, may well have been news released in September 1980 that New York gold prices had soared on reports that war had broken out.[15]

Most news reports from or about the war tended throughout to be discouraging. There were, however, occasional exceptions. The *Sunday Telegraph* carried one of these on 27 May 1984 from its Washington correspondent to the effect that the US and USSR had exchanged assurances that neither would intervene in the war.

Whatever the anxieties of the superpowers and the world's energy-consuming community, there were others whose concern was at least comparable, as they lived in the Gulf. The concern of the six states of the Gulf Co-operation Council (GCC) stemmed from several considerations: regret that two neighbouring Muslim states should be spilling one another's blood; anxiety lest the conflict provoke the intervention of one or both superpowers, with dangerous consequences; fear lest member states be attacked by one of the belligerents if they were suspected of sympathizing with the other; a feeling that a conflict of this sort diverted Arab and outside world attention and resources from the paramount issue of Palestine; a desire not to see either belligerent emerge from the conflict so strengthened by the spoils of victory as to become the most powerful entity in the Gulf; and, finally, anxiety lest their internal security should be threatened as a by-product either of the war or of Iran's revolution.

In addition, there was conceivably satisfaction that the war might divert Iran from any attempt to export its revolution southwards. The GCC six had been interested parties from the very beginning. Both Iran and Iraq had records of unpredictable behaviour, despite having maintained generally good relations with their neighbours along the Gulf's southern shore. During the Qasem regime in 1961 Iraq had laid claim to Kuwait in succession to Ottoman Turkey. Ten years later, Iran had occupied the three Gulf islands belonging to Sharjah and Ras al-Khaimah, having maintained for many years a claim to Bahrain (although this was dropped in 1970).

One important feature shared by most of the GCC states is an Iranian and/or Arab Shiite minority, whose behaviour in the event of some difference with Iran could not be predicted.

Of the six GCC members, Oman was the most directly involved, if only by virtue of geography.[16] It displayed particular concern over a possible closure of the Strait of Hormuz and a determination to take action to avoid this.

As the richest member of the GCC, Saudi Arabia inevitably commands most attention. It has a long record of allocating high proportions of its annual budgets[17] to defence, even appearing in the market for a submarine fleet in 1986.[18] This high defence spending continued as the war pursued its course: in 1988 defence and security claimed 36 per cent of the total budget.[19] This was made possible by the Kingdom's being the world's largest oil exporter and the possessor of its largest accumulation of reserves, estimated at some 165 billion bbl in 1980.[20] Saudi crude oil production over the decades preceding the war moved from 2.8 million b/d in 1967, via 8.4 million in 1974 to 9.9 million in 1980.[21] Revenues from crude production rose from $1.38 billion in 1967 to $108 billion in 1980. The GCC, however, was by no means a one-man band. Kuwait in particular, with its vigorous political and commercial life, had a powerful voice in GCC affairs. It had at one time been a larger crude producer than Saudi Arabia, with an output in 1964 of 2.3 million b/d against the Saudi 1.9 million b/d, only being overtaken in 1966.

Saudi Arabia had achieved a considerable degree of diversification in its defence suppliers. Since the 1960s both the US and Britain had supplied aircraft and related training programmes. On the ground, France matched the US as a source of battle tanks and surface/air (SA) missiles. It was nevertheless US equipment that was most in the news, after the struggle in Congress to carry through the sale of F-15 Eagles to Saudi Arabia in the face of resolute opposition from Israel and its supporters in the American legislature. The Saudis were also anxious to acquire their own AWACS aircraft, their own KC-10 in-flight refuelling aircraft and the Stinger SA missile, once again[22] a red rag to the Israeli bull.

Sensing a common concern over developments in the Gulf, the US offered (through Vice-President George Bush and Assistant Secretary of State Richard Murphy)[23] direct military assistance in protecting Gulf oil traffic. Saudi Arabia, supported by Kuwait and other GCC members, declined, suspecting that to agree would provoke internal tensions. The Saudis preferred to rely on their own defence capacity and that of their Arab allies. To reinforce this confidence, they had participated in GCC defence consultations a week or so before and joint manoeuvres with Kuwait the previous winter.[24]

One collective line of approach adopted by the GCC was to put forward a draft resolution to the UN Security Council, so worded as to highlight the Iranian role in endangering shipping, while remaining silent on that of Iraq.[25] In the final UN resolution, voted on 1 June 1984, the GCC text was somewhat watered down so that Iran's role was in less sharp focus. This joint initiative did not preclude individual mediation efforts by GCC member states:[26] in 1985 Saudi Arabia promoted an exchange of foreign ministers' visits with Iran and made other efforts to overcome the estrangement between the two countries resulting from the Eagle-Phantom air battle and Iran's resentment of Saudi and other GCC support for Iraq.

This support was to take two forms. A number of Arab states, adhering to the concept of Arab solidarity, provided moral or material aid for Iraq. Jordan, whatever King Hussein's private reservations about republican regimes in Baghdad since the assassination there of his cousin Faisal in 1958, provided its neighbour with port facilities at Aqaba and an overland route for supplies to Baghdad. Egypt, already a major source of manpower, contributed military spares with which, as a former customer of the USSR, it was handsomely stocked. Both Saudi Arabia and Kuwait, like Jordan, provided port facilities but their principal help was of a different kind.

The destruction in 1980 of Iraq's two Gulf terminals, together with the depletion of its foreign-exchange reserves, meant that it was in urgent need of an alternative means of delivering oil in the Gulf and a means of settling suppliers' bills while payment for the oil was coming through. The lion's share of both forms of help was to come from Saudi Arabia. Besides making crude available under various arrangements to enable Iraq to meet contractual commitments to deliver in the Gulf,[27] Riyadh entered into arrangements with Iraq to provide it with a means of moving its southern oil to the Red Sea.

Assessments of the scope of direct financial help for Iraq tend to vary. The Kuwait daily *al-Rai al-'Am*, for example, referred in 1981 to loans of some $12 billion (half from the Saudis, the rest split between Kuwait, Qatar and the UAE); the *Financial Times* in early 1984 put the Saudi contribution since the beginning of the war at $20 billion. Even in the absence of exact figures, this helping hand must be rated as 'very substantial'. Iraq, however, perhaps came to feel it could have been even more substantial. In 1984 *The Times* reported an interview with a Bahrain newspaper in which Foreign Minister Tareq Aziz[28] charged the Gulf Arab states with 'dragging their feet over help against Iran'.

This was not the way in which the Iranians came to perceive the Gulf states' role. In March 1986 a BBC monitor report told of demonstrations in Tehran by students and others, ostensibly Gulf Arab nationals, protesting against Kuwaiti and Saudi support for the Iraqi war effort.[29] Another BBC report at about the same time reflected official thinking: 'The wheel of Aflaqi[30] aggression has been turning with the help of Gulf petrodollars.'[31]

Shortly before this, the GCC foreign ministers had met in Riyadh and issued a communiqué: (a) 'drawing attention to the threats issued by Iran against member states [which] . . . created a climate of tension . . . helping to shake the security and stability of the region'; and (b) affirming 'the adherence of its member states to the policy of preserving the region's neutrality and resisting any action liable to carry the Gulf into the cycle of international rivalry'.[32]

Ultimately, whatever action the Gulf states took, Iran might choose to find fault with it. It was not surprising that Tehran had seized upon the state of the oil market, in the light of the widespread indictment of Riyadh as having provoked the circumstances which left OPEC in disarray in 1985–86.[33]

The BBC monitored Iran's comment. On 9 February 1986 Premier Moussavi was reported as saying that Saudi and Kuwaiti exports were the main factors in the collapse of prices. The oil minister, Gholamreza Aghazadeh, had declared the previous day that 'the US and its allies have been guiding certain oil producers in the "conspiracy" to bring prices down'.[34] Another major Iranian figure, Majles Speaker Rafsanjani, who occasionally functioned as a Friday Imam and whose *khutba* (sermon) was broadcast in Arabic on 28 February, pursued a similar theme:

> a new attack by Western imperialism in co-operation with reactionary agents . . . in order to support the criminal Saddam. The purpose of this attack is to weaken the economy of these [oil-producing and Islamic] countries and strengthen the foundations of the industrial countries' economies.

He persisted:

> All of us know that the Islamic revolution in Iran has caused the price of oil to triple. In spite of this, the new price equalled only half the real price of this vital commodity. Two years ago, while Iran was occupied in this all-out aggressive war, the colonialists began their moves to reduce oil prices. Regrettably, Saudi Arabia was the first to lead this campaign within OPEC.

(Then follow some remarks on Saudi-Iraqi co-operation.)

> All these actions led to the reduction of oil prices until they dropped to half of what they used to be . . .
>
> We expect the Islamic nation to support us in our latest struggle which we are carrying out in order to preserve the interests of the whole Islamic world.[35]

5
The Regional Dimension

Any review of the makings of this conflict must take account of the wider world in which the belligerents lived, of which some of the more important features were: the rivalry of the superpowers; the Arab-Israeli conflict; Islamic resurgence; the struggle for Arab world dominance; the world of oil and the emergence of OPEC; the Kurds; the political evolution of the Gulf states; Turkey and Afghanistan's position on the fringe of events; and the vicissitudes of the US dollar.

The Superpowers

The rivalry between the US and the USSR was sometimes direct and sometimes involved proxies. There is no shortage of information on this and the more general tensions between NATO and the Warsaw Pact, which all dated back to the immediate aftermath of the Second World War, when crisis succeeded crisis: Greece, Berlin, Korea, Cuba, Viet Nam, the Horn of Africa, Angola, Central America, Afghanistan. In the heart of the Middle East, there was superpower competition for influence in Egypt. Both the US and the USSR sent naval forces to manoeuvre about the Indian Ocean and both sought bases on the South Arabian coast. Both powers promoted defence sales to Gulf states, the US to Saudi Arabia, pre-revolutionary Iran and Oman, the USSR to Iraq and Kuwait. Both were prominent in world petroleum affairs and were major oil and gas producers; the US was also a net oil importer while the USSR was a former importer of Iranian gas and potentially interested in a resumption of this activity. Above all, each was determined not to allow the other to gain any strategic advantage or increase its general influence in the region. As both had their special preoccupations in 1980—the US its Tehran hostages and the USSR Afghanistan—there was little likelihood of Iraq's invasion provoking the intervention of either.

The Arab-Israeli Conflict

After the expiry of the Palestine mandate entrusted to Britain at the 1920 San Remo conference, the state of Israel, founded in May 1948 and since then cherished and sustained by Jews all over the world, entered the state of war with its Arab neighbours which has persisted ever since, further stimulated by the activities of Palestinian militants and their sympathizers both in the occupied territories[1] and farther afield. Both in the case of the PLO and its supporters and that of Israel and much of world Jewry, the degree of identification is reminiscent of that between a successful football team and its home crowd. Over the years since the end of the mandate, it is difficult to say which side, if either, has been the more intransigent. The persistence of this balance has placed the two sides in positions in which each is seeking something apparently unattainable.

Israel, now in possession of all the territory of geographical Palestine, seeks freedom from the threat of violence within the boundaries of this territory. The Palestinians, many of whom are desperate as a result of roots in the poverty of the refugee camps, hold few cards other than their determination not to be subdued and the 'armed struggle' through which this determination is sometimes expressed; what they seek is recognition of some form of statehood on Palestinian soil. Both wish Jerusalem to be their capital. All attempts to engineer a dialogue between the two have so far come to nothing, whether through UN Security Council resolutions in a sequel to the bilateral Camp David peace between Israel and Egypt in 1978, or through attempts to associate the PLO leadership with either Jordan or Egypt in the direct talks between them and the Israelis from which the latter have always recoiled. Back in 1980 there were no advance indications of the coming *intifadah* (uprising), followed by unprecedented signs of willingness on the part of the PLO to seek a negotiated settlement. Still less could it have been foreseen that Palestinians and Israelis would find themselves facing one another across a Madrid conference table in the 1991 encounter which provided most of the disappointments predicted by numerous observers at the time, but which may still prove to contain the seeds of a more fruitful sequel in the future.[2]

Resurgent Islam

Few countries with any sizeable Muslim population can have failed to experience some reflection of the mood of resurgent zeal abroad today in Islam. This zeal is projected in two directions. Inwards, it is expressed in a bid to lead the community in question back to a more meticulous observance of the canons of Islam according to which it is supposed to be living. This has resulted in the emergence of programmes of Islamic education or assertion, pressure within society for more modest dress for schoolgirls or women in general, opposition to mixed education, and so on. Outwards, there has been a call to overcome the sources of alien influence which are corrupting the Muslim societies with which they come into contact.

The appeal of this call is extremely powerful, partly for the sense of cleansing and exaltation it confers, but also, when actual conflict with the forces of darkness is involved, through the prospect of martyrdom. Martyrs (*shuhada*) who die in battle with the ungodly earn instant transmission to paradise (*al-Jannah*), the lush green gardens[3] from which comes the colour traditionally associated with Islam.

Apart from the role that this pattern of thought clearly played in shaping the Iranian revolution, there are a number of cases of its emerging elsewhere in recent years.

The Muslim Brotherhood first appeared in Egypt between the two world wars, as a reaction to the Europeanization of the urban bourgeoisie since the turn of the century. It called for a return to the *sunnah* and followed a tradition of Islamic resurgence in Egypt in the nineteenth century. The Brotherhood was suppressed by Nasser when he came to power in the 1950s, as it had been by his predecessors. One prime minister was assassinated by the Brothers; they failed in an attempt on Nasser himself. The Brotherhood is still active today, not only in Egypt where it is prominent in university life, but in Jordan and Syria (where it was blamed for the 1982 uprising at Hama). These exists a parallel women's movement, calling for a general return to modesty and the veil, in addition to the main programme of its male counterpart. Other Islamic movements have appeared in Egypt, such as Takfir wa Hijrah and the Gamaat al-Islamiyah. A fundamentalist group was charged with the assassination of President Sadat in 1981 and the atmosphere persisted throughout the 1980s.[4]

Similar movements have appeared elsewhere in North Africa: for example, the Tendance Islamique in Tunisia, the Front Islamique du Salut (FIS) in Algeria[5] and the Green Revolution in Libya, which is now virtually the official programme of the Qaddafi regime and widely identified with the Colonel personally.

In Lebanon, where the overall Muslim/Christian demographic balance was historically translated into a 5:6 ratio in parliamentary seats under the 1943 National Pact, Muslim efforts at self-assertion were initially directed, without particular differentiation between Sunni and Shiite, towards overcoming Christian domination of the entire political system. In the 1980s both sects developed resurgent movements. The Sunni version was Islamic Tawheed, which was particularly strong in Tripoli. The Shiite-generated Amal and Hezbollah, underpinned by Syria and Iran respectively, were inclined to sometimes violent disagreement with one another.

Turkey had its Islamic groups too. Ever since the Second World War, there had been pressure in some circles for a move away from the secular aspects of the state created between the wars by Ataturk. This centred originally on the right-wing Nation Party and the sentiment has persisted, with clashes between rival fundamentalist groups common before the 1980 military coup.

In the Gulf, there was the spectacular armed seizure of the Mecca shrine late in 1979 by a band of zealots of mixed nationality claiming to include the Mahdi, or 'divinely guided one', whose appearance Muslims await. The extent of popular support, if any, for this enterprise was never clear. Certainly none was evident in Riyadh.[6] At about the same time, in Saudi Arabia's Eastern Province, there was unrest among the Shiite minority, though not obviously connected with the Mecca incident.

The Struggle for Arab World Dominance

There have been a number of competitors for the dominant role in the Arab world. From the 1950s Egypt sought, with the prestige gained by Nasser's success in severing the residual links with Britain, to achieve primacy in three distinct political systems: post-colonial Africa, the Arab world and the Islamic world.[7] In this it occasionally ran up against other states with their own sense of destiny. Syria, drawn into partnership with Egypt in the United Arab Republic created in 1958, withdrew from it in 1961, and at different times the two crossed swords through proxies in Lebanon. Saudi Arabia, devoted to the twin concepts of Islamic and Arab solidarity, was for many years active in promoting both, aided by the twofold prestige of guardianship of the Holy Places in the Hedjaz and the country's immense petroleum and financial reserves. Saudi Arabia and Egypt clashed through the parties they respectively supported in the Yemen civil war in the 1960s.

Other states that have given the impression of having comparable senses of destiny are Libya and Iraq. Even before the Iran-Iraq war began in 1980, Iraq had always taken an active role in Gulf affairs and those of the Arab world in general, especially in relation to Palestine. Libya, despite its geographical remoteness from the heart of the Arab lands, shared Iraq's concern to be involved in deciding crucial questions affecting the Arab world.

The World of Oil and the Emergence of OPEC

Any modern account of oil in the Middle East has to begin with OPEC. The Organization of Petroleum Exporting Countries came into existence in Baghdad in 1960. It was intended as a device to enable states with major accumulations of crude oil to exercise the fullest possible degree of control over the industry concerned with its extraction. The typical picture in such industries was one in which international oil companies, having secured concessions under which they risked their capital, discovered and developed the resource. They then set the price of oil at its loading port and found outlets for it in world markets. In the process, as OPEC saw it, the oil companies, together with the governments of consuming countries, realized far too large a share of the available economic rent,[8] the margin between the cost of producing the oil and the price it fetches from the end-consumer. Since the companies set the price, they determined both the amount of profit per bbl from extracting it and the amount of tax the 'host government' could levy on the process. This was a state of affairs that OPEC was determined to change.

After an initial success in 1964 over 'royalty expensing', an element in the intricate mechanics of computing concessionaires' tax liability which gained OPEC members some extra cents of revenue per bbl, they moved on to the earnestly disputed negotiating rounds in Tripoli and Tehran in 1971. Substantial improvements were gained before OPEC moved on the following year to attack the issue of equity participation by host governments in the businesses engaged in extracting and exporting the mineral. These they now all control.

During the 1973 Arab-Israeli war Arab petroleum-exporting countries, banded together as the Organization of Arab Petroleum Exporting Countries (OAPEC), engineered destination boycotts of the US and Netherlands. The following year, OPEC's Tehran conference brought negotiations with the oil companies to an end, with legislation henceforth the order of the day. Prices were on the rise, having trebled in 12 months, and were now entirely the province of the producer-country governments. This and other changes that year made an important contribution to the wave of inflation which beset the oil-consuming world for the rest of the 1970s and led to the formation of IEA towards the end of 1974.[9]

The Kurds

The Kurds see themselves as the forgotten people of modern times, although few have forgotten the name of the best-known Kurd of all—Salahuddin or Saladin. Some Kurds believe their race descends from the Medes of early Persian history. Today they are spread across an area overlapped by Iran, Iraq, Turkey, the USSR and parts of Syria. The 1920 Treaty of Sèvres, part of the array of accords reached by the Allies at Versailles after the First World War, encouraged the Kurds like the Armenians to hope that their aspirations for some form of statehood might be fulfilled when the Ottoman empire was dismembered.

No such state emerged, however, and 1945 found them no farther along the road than they had been in 1918. An attempt was made from 1945 to 1946 with the inauguration in northwest Iran of a Kurdish republic[10] centred on Mahabad, but after a brief existence it was suppressed and its president executed. At about that time, Kurds in Iraq, hoping perhaps to emulate the positive side of what had been achieved over the border, rose in revolt. This and later attempts led nowhere; insurgency persisted into the 1960s. Although a settlement was negotiated in 1970, there were disputes over its implementation and troubles continued throughout the 1970s and 1980s. Iran's support for such activity was, as previously mentioned, one cause of the poor relations between Iraq and Iran overcome at Algiers in 1975. Kurdish insurgent activity was not confined to Iraq and Iran but contributed to the atmosphere of insecurity that prevailed in Turkey before the military coup in 1980.

The Gulf States

The Arab states of the Gulf went through an eventful spell of political change in the 1970s and 1980s. Wearied by the complications of its withdrawal from Aden, in 1968 the British government announced its decision to disengage east of Suez by 1971. This gave rise to a scheme to incorporate into a federation all the previously British-protected emirates in the Gulf south of Kuwait. Bahrain and Qatar eventually decided not to join and the UAE came into being, now a substantial oil producer and member of OPEC.

In 1981 a wider umbrella organization was set up—the Gulf Co-operation Council (GCC), which grouped together the UAE, Bahrain, Qatar, Oman, Kuwait and Saudi Arabia.[11] An initial preoccupation was co-operation over economic development, but later priorities were internal security and defence.

The States of the Northern Tier

The year 1980 was also important for other states in the 'arc of crisis' surrounding the lucrative petroleum province of the Gulf. In Turkey, the September coup[12] brought to power a military regime headed by General Evren, cutting short the period of civilian parliamentary politics in which the personalities of Ecevit and Demirel were prominent and which was marred by much violence, urban and rural alike, with right- and left-wing extremists involved.

Farther round the arc was the Soviet incursion into Afghanistan, which began in 1979[13] and lasted for most of 10 years, despite resistance by the Mujahedin. One noteworthy international impact of this episode in 1980, following the ultimatum from President Carter, was the US boycott of the Moscow Olympic Games.

Finally, in the Indian sub-continent, India was involved in a close relationship with the USSR, and Pakistan with the US and also with the People's Republic of China.

The Fortunes of the Dollar

One of the issues raised by OPEC, in intervals between efforts to raise the unit price of crude oil, was the periodic decline in the value of the currency in which all crude oil trading is conducted, and in which therefore OPEC members' revenues were received. This decline was often caused by US balance of payments difficulties or the mechanics of US budgetary management. Both adverse US payments figures and US budget deficits are familiar stimuli to dealers in dollars, as are hints of any prospect of the money markets' interest rates being raised (a classic central banker's response to market pressure on the currency in which the bank operates).

The 1970s witnessed a number of developments of this sort[14] and during the days when OPEC was still negotiating with the companies over prices, a device was contrived whereby if the dollar went down, the oil price went up and vice versa. This arrangement was reached at two rounds of negotiations at Geneva, designated by the press contingent Geneva I and Geneva II.[15] The device ended on a sour note from the consumer standpoint in 1974, however, when the dollar underwent an increase in value which ought to have triggered a price reduction but somehow failed to do so.[16] Geneva I and Geneva II were not much talked about thereafter by either side.

During the 1980s the fortunes of the dollar hovered ominously on the fringe of events. In 1986, when oil prices weakened, an already serious US trade deficit was aggravated by the effect of low-priced oil imports which 'sparked nervous trading on currency markets'. The figures were seen as evidence that both the dollar and US interest rates should drop to correct the huge trade imbalance which was projected at a record level of about $165 billion for the year.[17]

Map II. The Inland Battlefront

6
The Inland Battlefront, 1981–88

The initial exchanges of the war left both sides with some damaged oil installations and Iraq still in possession of a limited amount of Iranian territory. Iran's main riposte to the original Iraqi push took place in the south and centre of the 730-mile (1,175 km) long battlefront to which both sides were to become accustomed. It was designed to rid Iran of the remaining elements of the invading force in these sectors (some having withdrawn of their own accord) and took the form of powerful punches to which distinctive names were generally given, evocative of important features of Islam's past.

The first of these, aimed at breaking the siege of Abadan, was Thamen al-Aemmah,[1] mounted on 27 September 1981. Between November 1981 and November 1982 there were four more:[2] the first, Tariq al-Qods, was launched near Bostan and the last, Bait al-Moqaddas, was close to Basrah and Khorramshahr. A further series followed in the latter half of 1982.[3] Then in 1983 and 1984 came the series of attacks to which Iran gave the name Val Fajr (The Dawn).[4] These occurred at different times down the whole length of the front, from Haj Omran in the mountains of the north to the marshes in the south. During these years two important features of the war—one on each side—came into focus. The Iraqis, once they had withdrawn from the territory occupied in the early months, and to some extent even before this took place, constructed solid defensive positions, notably in the centre near Mandali and Khanaqin, and in the south close to Basrah, from which to fight off the anticipated Iranian offensives. In the case of the Iranians, early in 1982 came the first references[5] to the phenomena of attrition and the 'human wave'.

The South

This sector was to see the bloodiest encounters—and some of the most grisly reporting—of the entire war.[6] It contained the most unpromising terrain for modern warfare of the whole extent of the battlefront—the habitat of the Marsh Arabs,[7] the cluster of lakes north of Basrah, particularly ill-adapted to the deployment of battle tanks and other tracked vehicles.[8] One dispatch from this front during an Iraqi attempt to recover the captured Majnun oilfield was headed 'Thunderstorm forces pause in Gulf war'.[9] There was nothing new about this experience; Mesopotamia was discovered to be unfriendly terrain for military manoeuvre during the First World War[10] and British service maps of the area are peppered with such legends as 'motorable in dry weather', 'flats liable to flood' and 'marshland usually flooded'.[11] The Iraqis made the fullest use of these defensive advantages in 1984 to supplement the positions they had prepared on dry land while awaiting the half million or so Iranians whom US satellite photographs had identified as massing on the frontier near Basrah.[12] It was not quite a desert Maginot line, but a solid enough substitute, with the artillery component sometimes consisting of battle tanks sitting hull-down in specially excavated 'scrapes'.[13] To supplement the natural barrier provided by the marshes themselves, ditches were constructed that could be flooded[14] with giant pumps[15] and even electrified,[16] to create what was described by the *Guardian*'s correspondent as a 'giant moat'[17] and by the Iraqi corps commander of the area[18] as a 'fish trap'.

For all the unmanoeuvrability of armour in the inhospitable southern sector, the Iraqi leadership evidently decided to retain the option to bring it into play at some stage, at least during the warmer months. There were reports of their best equipment, the Russian T-72, being pushed south on tank transporters along the north-south axis.[19] Another card the Iraqis kept up their sleeve was the helicopter gunship.[20]

One of the most dramatic developments in the southern sector during the first half of the war was Iran's capture in February 1984 of the Majnun oilfield[21] and its recapture, if only partial, in June 1985.[22] In between, there were a number of clashes during which Iraq attempted to retake the area.[23] By 1988 it was reported to have been fully recovered.

Iran's strategic aims in all these attacks appear to have been, first, to take Basrah and, second, to cut the strategic road system linking it with Baghdad. One of these involved particularly fierce fighting during an assault on the villages of al-Beidah and al-Sakhrah, for which the Iranians used both boats and helicopters to bring in their troops. The Iraqi commander, Lieutenant-General Maher Abd al-Rashid, afterwards denied charges that Iraq had used mustard gas in this engagement.[24] Fighting of this sort was to continue in the marsh area, but it was not until 1986 that Iraq was able to recover more of Majnun.[25]

The Centre

This stretch of frontier, at which part of Iraq's original 1980 offensive was directed, was one to which both sides attached a good deal of importance. This was principally because it contained the old Baghdad-Tehran highway. The Iranians saw domination of this hilly area as throwing the way open for a push towards Baghdad. For the Iraqis, the appeal was mainly the converse: it was a means of pushing armour, if not to the outskirts of Tehran itself, at least towards other important targets on the way, such as Kermanshah (now renamed Bakhtaran) and Hamadan. On the Iraqi side of the frontier, the town of al-Amarah, on the easternmost of the three Baghdad-Basrah roads, was seldom free of military activity, with convoys of trucks and armour pouring into it, and private cars and taxis carrying battlefield coffins on their roofs frequently setting off in the opposite direction. One rainy afternoon in November 1982, when I was returning from Basrah to Baghdad by car, we were overtaken and swept along by the motorized column of President Saddam himself. As emerged later from the television news, he had been visiting the front near al-Amarah, during what must have been part of the offensive named by Tehran as Muharram al-Haram.

Attacks at or near this point continued as the war went on. Another came in 1984 in the course of Val Fajr No 6. One objective of this push, and the second part of the same offensive farther down the Tigris at its confluence with the Euphrates, was to cut the Baghdad-Basrah road. Iran claimed to have liberated 37 villages[26] in the provinces of al-Amarah and Basrah, but as far as the front was concerned, Iraq's defences near al-Amarah were reported to have held. Further attacks near al-Amarah came in mid-1985 with the offensives named al-Qods (Jerusalem) Nos 2 and 4.

While al-Amarah remained an exceptionally lively point in the central sector, as the 1980s advanced there was more activity up towards the northern end, near Mehran. Early in 1984 the Iranians launched Val Fajr No 5 between Mehran and Dehloran.[27] They followed this a few days later by a two-pronged push, one prong spearing towards the Bostan-al-Amarah area and the other towards Ali al-Gharbi on the Baghdad-Basrah road, meeting some success in occupying heights overlooking the town.

Two years later, Mehran was still the scene of a good deal of activity. In early May, Iraq succeeded in taking the town, the first move into Iranian territory since 1982. On 1 July it was recaptured by Iran, in the course of the offensive[28] named Karbala No 1. After a further two months, Iran conducted another advance west of Mehran, which Baghdad duly reported as having been crushed.[29] Not far away was Sumar, west of Bakhtaran, where Iran launched a major offensive called Karbala No 5 in January 1987. Finally, in July the following year, it was time for an assault in the opposite direction when forces of the National Liberation Army of Iran[30] (in fact, the Mujahedin-e-Khalq, led by Rajavi) entered Iran with Iraqi support, west of Eslamabad (previously Shahabad).

The North

Whereas the central sector is mainly hilly terrain and the south is flat and marshy, the north is the only part of the frontier in which real mountains are to be found. Iraq's Kirkuk oilfield lies in the part of the northern sector closest to the capital, in an extension into Iraq of the Zagros mountain chain which dominates western Iran. The greater part of this sector on both sides of the frontier is inhabited by Kurds.

In Iran, this minority was regarded by the new regime with, if anything, less affection than the old,[31] which had crushed the short-lived postwar Mahabad republic. In recent years, the banner of Kurdish separatism has been borne by the Kurdish Democratic Party of Iran (KDPI), led by Dr Abdul Rahman Qassemlou until his assassination in Vienna in 1989.[32] In

Iraq, the Kurdish Democratic Party (KDP) was led first by Mullah Mustafa Barzani, a survivor of the ill-fated Mahabad venture in Iran, and after his death in 1979, by his sons Mas'ud and Idris. By a curiously symmetrical piece of political manoeuvring, the KDPI, in revolt against Tehran, was succoured by Baghdad, while the KDP, in revolt against Baghdad, was succoured by Tehran.

The KDP was not alone in Kurdestan in opposing Baghdad; there existed another Kurdish separatist group in Iraq, the Patriotic Union of Kurdestan (PUK), led by Jalal Talabani. The KDP was dominant in the far north, its main centre being Haj Omran beyond Mosul, while the PUK's main support came from around Suleimaniyah, nearer to Kirkuk.

The KDP were accustomed to retreating across the frontier into Iran and Idris Barzani died there in 1987.[33] The KDPI, on the other hand, would make their way across the frontier into the Suleimaniyah area. In 1985 their Peshmerga (guerrilla fighters) were encamped here under Baghdad's protection while not far away were the still-insurgent Peshmerga of the PUK. The KDPI were able to come and go across the frontier and through the no man's land between the opposing armies with great freedom.

One witness to this intriguing feature of the northern front was a British surgeon. Dr Nicholas Parkhouse had come to Kurdestan in 1985 under arrangements set up by International Medical Relief, a body linked to the French Médecins sans Frontières which was organizing a medical assistance operation spanning both sides of the frontier at the request of KDPI leader Qassemlou. An interesting sidelight on battlefront conditions came from his remarks that the Peshmerga brought in for attention mainly tended to be suffering from wounds to their limbs from weapons such as Kalashnikovs, since those who sustained head or body injuries generally failed to get as far as a field hospital.

Fighting in the north was seldom on the same massive scale as in the centre and south, although this sector had its share of offensives as mountaintops commanding important centres like Suleimaniyah and Haj Omran changed hands. The peak that featured most often in battlefront reporting was Jabal Kardmend.[34] Another geographical feature often in the news was the Panjwin salient, a narrow tongue of territory stretching out into Iran from east of Suleimaniyah, attacked by Iran in Val Fajr No 4 in October 1983.

The PUK had not always been devoid of outside help, having at one stage been supported by Syria[35] before entering into negotiations with

Baghdad in 1984 for a new deal for Iraqi Kurds to improve on the Autonomous Region status conceded them under a mosaic of earlier legislation. These negotiations did not go well, however,[36] and by 1985 the PUK were again in rebellion.[37] Although the KDP were also in revolt at the same time, there were too many differences between them for there to be any prospect of co-ordinated action. There has at times been a degree of common cause between the PUK and the Iraqi Communist Party (ICP), which was once involved in a national front with the Baath, but subject to repression in recent years. This common cause blossomed to embrace other anti-regime groups and by 1988 efforts were being made in Damascus to assemble an anti-Baghdad coalition under ICP leadership.

During this period, the most effective co-operation concerning the Kurds was probably that between their adversaries, in other words, the governments of Iraq and Turkey. Turkish troops conducting operations against Turkey's own Kurds (among whom the left-wing Kurdish Workers' Party, PKK, were especially active) were accorded the right to cross into Iraq[38] in hot pursuit.

Returning to Iraq's frontier with Iran, and to the part of it closest to the PUK heartland around Suleimaniyah, we may now look a little more closely at Val Fajr No 4. This offensive took place between Marivan and Sardasht on the night of 19–20 October 1983[39] and may have been intended to cut the lines of communication between the Iraqi forces and the KDPI. Some three months earlier, Iranian forces with KDP support had made gains near Haj Omran.

A much larger-scale offensive, codenamed Tahrir al-Qods (Liberation of Jerusalem), came on 12 February 1984, this time south of Marivan, towards Darbandekhan,[40] southeast of Suleimaniyah. It was widely believed that, as Iran was known to have massed troops all the way along the frontier, this move, though sizeable, was a trick to conceal the real intention, a thrust in the south or centre. In the event, after Iran had claimed the capture of 40 square miles (100 sq. km) of Iraqi territory in the north, this was followed by the next attacks due in the Val Fajr series, Nos 5 and 6, both in the centre, and Val Fajr No 6 partly in the south. They were described as mere preparations for an Iranian final offensive,[41] though this much-heralded event was to be a long time coming.[42]

Fighting nevertheless continued in the north, occasionally varying in intensity. February 1986 saw Val Fajr No 9, during which Tehran claimed the capture of a number of important heights,[43] followed by a major Iraqi sweeping-up operation in the Panjwin area[44] the following month. Then in September came Karbala No 2, an Iranian attempt to recapture Jabal Kardmend, lost the previous May.[45]

Val Fajr No 9 was to be the last set-piece operation in the north.[46] On 7 March Tehran radio claimed that Suleimaniyah and Chawarta were both overlooked.[47] A few days later, Baghdad announced the liberation of three strategic heights[48] by the First Army Corps. After these exchanges, fighting in Kurdestan went on for some considerable time. Moreover when the Iraqi First Corps swept up at Panjwin, they evicted Iranian forces that had been in possession since 1983.[49]

The final throes of fighting in the north came in 1988, but this time the Iranians were not involved directly. In February, just as Kurdish leaders were predicting a final Iranian offensive on all three fronts, the Kurds stepped up their pressure on what they perceived as Iraq's weak spot in their own area, claiming a victory over government forces at Deeralok,[50] near Amadyah. During the following two months came the first reports of Baghdad's retaliation at the frontier town of Halabjah,[51] in PUK territory, where aircraft were said to have dropped bombs containing poison gas.[52] Many Kurds were found to be dead, variously reported as counted in hundreds and thousands. Inquiries were mounted by the UN and Amnesty International. Despite some initial uncertainty as to whether these accounts were correct, it became generally accepted (e.g. by the US and British governments) that chemical weapons had indeed been used by Iraq on the scale reported. Later thousands of Kurdish refugees sought sanctuary in Turkey[53] and Iran.[54]

Success for Iran

After so many offensives with mixed outcomes, fortune finally smiled on Iran in 1986. This change came with the eighth in the Val Fajr series, launched in the south on 9 February in commemoration of the period seven years earlier when the Shah left and Khomeini returned from his sojourn in France. By all accounts displaying considerable skill, the pathfinders of the assault made their way across the Shatt al-Arab to land on the Fao peninsula, occupying the island of Umm al-Rasas on the way over. Once ashore, the attacking force made for and took the remains of the Fao oil-export complex.[55] Having achieved this, they provoked disquiet in two quarters:[56] first, those concerned with the defence of Basrah, up the river to the north, and the nearby port of Umm Qasr; and, second, the government of Kuwait (an important source of support for Iraq throughout the war), which had already suffered air attacks and was to be struck again before the fighting was over.

The Iranian forces at Fao were finally unable to reach either Basrah or Kuwait. The Iraqis moved armour into positions south of Basrah to block the more obvious of these steps and the Iranians evidently decided that Kuwait could be attended to some other time. They did, however, turn towards Umm Qasr, up the Khor Abdallah,[57] but were brought to a halt in circumstances which are still unclear, Tehran[58] accusing Baghdad of having used chemical weapons.

The fighting continued in the peninsula for some weeks, with Tehran radio announcing attacks by the 'army of monotheism' along the Fao/Umm Qasr and Fao/Basrah axes.[59] Meanwhile Iran had achieved a considerable boost to its morale: many foreign commentators believed that, with this achievement under their belts and the onus on Iraq to winkle the Iranians out of Fao, the war was now going Iran's way.

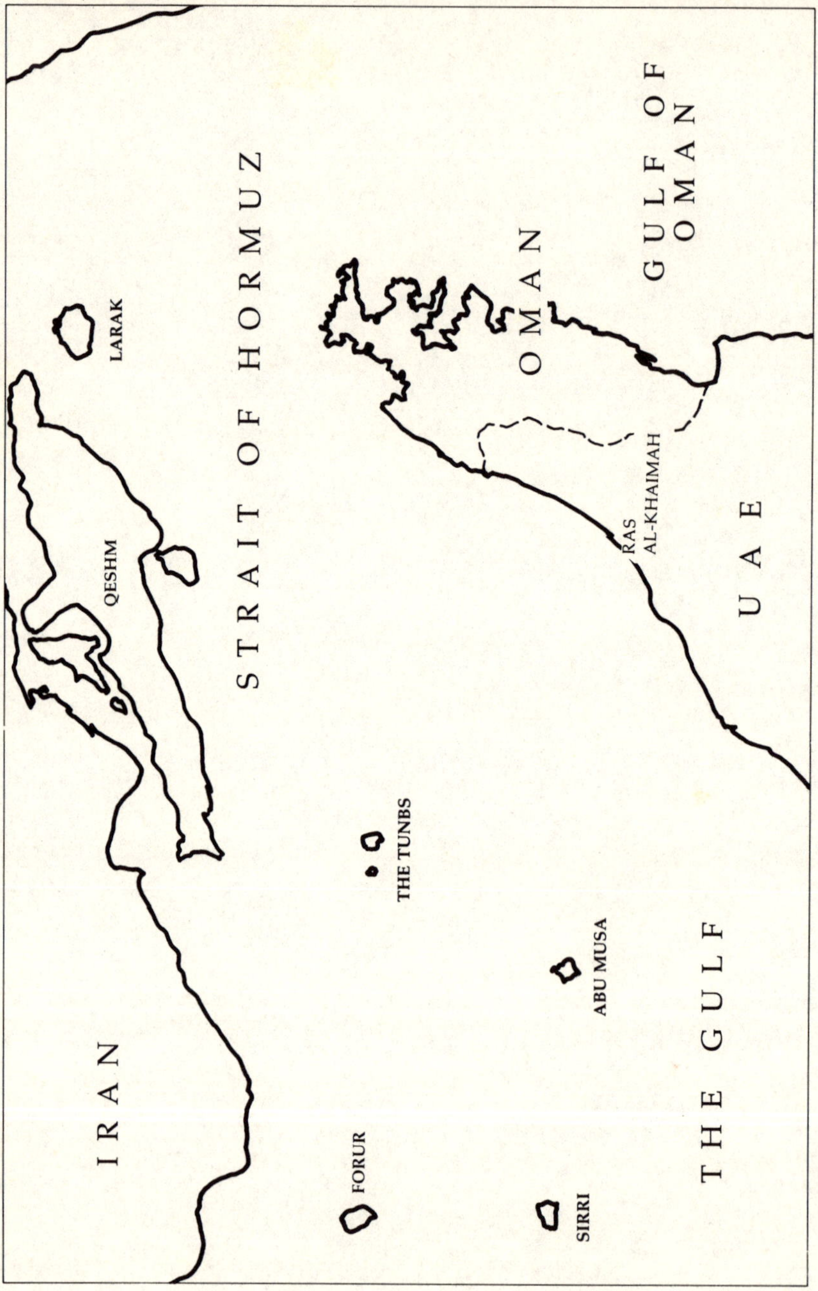

Map III. The Entrance to the Gulf

7
The War at Sea

One respect in which the war came to acquire an international dimension concerned its impact on shipping. It was in 1984 that the war at sea began to hit the headlines, although there had been earlier incidents involving attacks on shipping. The first vessel casualty recorded was the bulk carrier *Louise I*, hit and sunk in May 1981 in an Iraqi air attack.[1] But if 1984 was the year in which the expression 'tanker war' was coined, 1987 was to prove the most eventful of the entire conflict.

In May of that year there was an Iraqi Exocet missile attack in the Strait of Hormuz on the USS *Stark*, which was engaged in the protection of merchant shipping. The news broke after something of a lull in the abundant reporting of incidents from the various scenes of fighting. Iraq subsequently claimed that the attack was in error, but the news came as a considerable surprise both to the inhabitants of the Gulf and to professional 'Gulf-watchers' from outside the area.

This episode, which cost the lives of 37 seamen, was significant in five further ways. First, it revived international concern over what had often been called the 'forgotten war'—although it had by then lasted longer than the Second World War, it was still far from being brought to a formal end. Next, it underlined the dismay with which many observers viewed the prospect of direct superpower involvement in Gulf affairs. Third, there was much concern over the continuity of oil and gas movements out of the Gulf area, on which a host of consumers, in the Third World as well as OECD countries, depended. Fourth, there was anxiety over the extent to which the Gulf states, whether formal belligerents or not, had been arming themselves with the most advanced weaponry available from a wide range of sources. Finally, it served as a timely reminder of how dangerous navigation in the Gulf had become since fighting broke out in earnest in 1980.

The war at sea was provoked by each side's determination to deny the other the means of exporting its oil. The Iranian attacks in September 1980 on Iraq's southern terminals put them totally out of action.

Recognizing the danger of Iraq's retaliating against Iranian terminals, Iran decided from the outset to discourage foreign incursions of any sort. It issued a warning to all shipping to stay outside its territorial waters (now extended to a distance of 12 miles (19 km) from its coastline) and more generally outside the 'Iranian half' of the Gulf[2] up to the median line between the two coasts. This had the effect of realigning the traditional shipping lanes up and down the Gulf which led to the centuries-old general cargo ports in the Shatt al-Arab, Basrah and Khorramshahr, as well as the old-established Gulf ports like Bahrain and Kuwait and the recently installed array of oil terminals nearby: Mina al-Ahmadi and Mina Abdallah in Kuwait; Ras Tanura and al-Jubail in Saudi Arabia; Mina Saud and Ras Khafji in the Neutral Zone between the two; Sitra in Bahrain; Halul Island off Qatar and its onshore counterpart Umm Said; Jabal Dhanna in Abu Dhabi and offshore Das Island; and Kharg Island and Bandar Mah Shahr on the Iranian coast. These were the main sources of oil and gas in the Gulf, though there existed a handful of others.

Oil consumers, whether directly or indirectly concerned, concentrated their attention on the Strait of Hormuz at the mouth of the Gulf, through which flowed some 20 per cent of the oil supplies of the world outside the CPEs[3] and which shifted into sharp focus as the war intensified. Not all the Gulf output was at risk. The initial outbreak of war had already taken around 3 million b/d out of the world supply/demand equation. As Saudi Oil Minister Yamani later remarked,[4] this had the one beneficial effect of diminishing a prevailing market surplus that year. Because Saudi Arabia had an alternative outlet on the Red Sea at Yanbu', created in the 1970s against just such a contingency, it was estimated that only some 8 million b/d would be removed from the market were Hormuz closed. It seemed likely that about half this shortfall could be made up from other sources, some within OPEC and some outside it (Mexico, the North Sea and others).

One other consequence of the closure of the strait, or the threat of it—foreseen by Shaikh Yamani, in something of an aside—was that Lloyd's might cancel insurance for the Gulf. The oil minister suggested that this would be as serious as physical closure, in which case 'we are heading for real trouble'.[5]

The perception of the consumer world (IEA, its individual member governments, oil companies, specialist observers and interest groups) of a possible closure of the Gulf was based on the same data as those underlying the Saudi analysis. From the consumer standpoint there were two main hazards. First, there was the risk of physical shortage, causing dislocation and possible hardship. Second, there was the threat of a repeat performance of 1979, when the evaporation of 5 per cent of crude oil availability had brought about a doubling of prices.

Against these spectres the OECD/IEA world (without in all cases necessarily seeking to act collectively) could envisage two possible lines of action: (a) setting in motion the IEA contingency mechanism for an 'emergency sharing program';[6] and (b) taking military action (e.g. on the lines of the 'Carter doctrine') to preempt the development of a shortage, in other words, preventing any breakdown in navigation. Option (b) could have taken the following forms: (i) an attack on the forces/territory of those seeking to obstruct free navigation; (ii) escort activity to defend merchant shipping seeking to exercise the right to free navigation; and (iii) clearing wreckage or mines endangering free passage through Gulf waterways like the Strait of Hormuz.

Of these, (i) was on the face of it within the scope of the US Navy task force deployed in the Indian Ocean and led by an aircraft carrier.[7] In practice, the only action undertaken was a combination, in conjunction with European navies, of (ii) and (iii) in 1987. There were, however, earlier reports of US Navy craft (using their docking facilities in Bahrain) moving behind tankers making their way through the Gulf by night, keeping an unsolicited eye[8] on them.

Well before the operations of 1987, the prospect of foreign intervention of any kind was arousing concern in several quarters. One of these was the GCC. A declaration by Saudi Defence Minister Prince Sultan ibn Abd al-Aziz on a visit to Qatar in 1984 probably reflected the atmosphere of GCC deliberations at the time. 'The sons of the Gulf', he said, 'are capable of defending their region against any aggression.'[9]

Another key factor related to the possibility of the closure of Hormuz was the mood prevailing in ruling circles in Tehran. With a ground offensive widely expected in 1984, Iran's President Ali Khamenei described closure as an option his country was indeed contemplating. He accused the 'supercilious world powers' of plundering the oil resources of producing countries:

> Oil must be used as a weapon against the common enemies of the Muslims. The closure of the Strait of Hormuz will bring to a stop the wheels of Western industries. We now hold this weapon.[10]

Further limitation on freedom of manoeuvre, this time well up the Gulf, was introduced by Iraq. This may have been prompted by its loss of oil-export flexibility, caused first by Iranian naval action in 1980, then compounded by Syria's pipeline closure two years later. In August 1982 Baghdad announced a military exclusion zone at the head of the Gulf, covering the oil terminal at Kharg, the general cargo ports of Bushehr and Bandar Khomeini and their respective approaches. These dispositions and their object—that of bringing Iran's crude-oil export system to a standstill—were set out with ample detail and forceful purpose[11] by the country's leadership.

Bandar Khomeini and Kharg were subsequently attacked by Iraqi aircraft and hits on shipping claimed. Kharg itself was reported as solidly defended, with Raytheon Hawk surface/air missiles[12] and F-14 fighters operational in the area,[13] and it was some time before significant damage was inflicted on the terminal.

Another problem to surface in 1982 was the reluctance of Japanese seamen to man vessels bound for the Gulf. This attitude fluctuated as work resumed,[14] but five years later, Japanese seamen were again[15] boycotting the northern end of the Gulf. Over succeeding years, the seafarers' unions of several countries managed to secure improved conditions for work in the area.[16] Norwegian seamen who entered the Gulf received triple pay, for example. In 1987 Greek seafarers won a special deal for Gulf service, with allowances of up to 140 per cent on basic wages. A German captain, interviewed off Larak Island by a *Times* correspondent, acknowledged the risks involved but went on to point out that 'the money is good and the contract is only six months'.[17] In 1983, despite the evident risks, and with claims that Kharg Island had been destroyed at least once,[18] loadings continued. Fleet operators appear to have felt that the raids had not had much effect on operations there and it was rumoured that the National Iranian Oil Company (NIOC) was managing to improvise, using storage tanks on the mainland. The storage problem persisted until 1986[19] and may have contributed to NIOC's later decision to bypass Kharg altogether.

A further hazard with which seafarers had to contend was mines. When the 'reflagged' Kuwaiti Ultra Large Crude Carrier (ULCC) *Bridgeton* struck a mine in the Gulf while sailing in a US naval convoy in July 1987,[20] this came as no surprise to those familiar with recent developments in the waterway. A Russian tanker had struck one off Kuwait earlier in the year.[21] In July 1987 US Navy divers discovered 10 mines off Mina al-Ahmadi. They were clearly not 'drifters', as they were found planted on the sea bottom in the port's approach channel.[22]

Mines had been a potential hazard at or near the scene of fighting or port approaches since 1983. One had been found that year floating off the coast of Qatar, prompting search operations by GCC states up and down the Gulf.[23] Then an Iraqi spokesman claimed that Iraq had mined the approach to Bandar Khomeini,[24] though this was contradicted by Iran. By August 1984 a report issued by Lloyd's of London was identifying mines, together with missiles and groundings, as the principal causes of damage claims.[25] Back in 1983, one exploit that can with reasonable certainty be attributed to Iraq was the attack in July on Iran's small Nowruz offshore oilfield, causing a large crude-oil slick and heightened environmental concern[26] around the Gulf.

Probably the most dramatic development of 1983, and one which foreshadowed the USS *Stark* episode, was the arrival in Iraq in October of five French Super Etendard aircraft, complete with Exocet missiles.[27] However, the aircraft that caught the *Stark* was a French Mirage.

The year 1984 was particularly eventful. It was the first in which a substantial number of attacks on tankers occurred, 62 being reported in all.[28] The first of these was the Greek tanker *Filikon L*,[29] reported to have been attacked by an Iraqi aircraft. The next five craft to be attacked were also generally attributed to Iraq. Not all were tankers—the second, a South Korean-run oilfield service vessel, was caught close to a Saudi offshore field and sunk.[30] In April two Saudi-flag tankers, one the ULCC *Safina al-Arab*, appear to have been caught quite separately by Iraqi aircraft close to Kharg. It later emerged[31] that they had loaded crude here for third parties to whom they had been chartered-out.

After Iraq had hit another 2 tankers, 1 Iranian and 1 Panamanian (making 7 out of the year's total of 44), it was Iran's turn. The target selected indicated a new development. It was the Kuwaiti-flag *Umm Casbah*,[32] carrying Kuwaiti oil to Britain for Shell. By the time Iran had attacked a Kuwaiti, a Saudi and then a Liberian tanker (the last of these caught near the principal Saudi terminal at Ras Tanura), the GCC states had begun to recognize this sequence—as it was doubtless intended—as a signal to distance themselves from Iraq in the light of the support they had been providing for Baghdad.[33] In spite of further such signals as the war progressed,[34] help to Iraq by GCC members continued.

The next flag to be attacked was that of another friend of Iraq, though this time the attack came from Iraq itself: Turkey's *Buyuk Hun*[35] was taken near Kharg. Shortly afterwards came a Saudi response to Iran's continuing attempts to teach Riyadh a lesson. On 5 June one of two Iranian F-4s that had come over from Bushehr, looking for another kill off the Saudi coast, was shot down by the Saudis' more advanced F-15s, armed with the Raytheon Sparrow missile and guided on to their targets by AWACS surveillance craft.[36]

Two days previously, Iran had announced a no-go 'Advice Area' of its own,[37] running from a point close to Kharg and across to Kuwait, down the Gulf to the islands of Sirri and Abu Musa.

Oil companies were in the news that year. A group of American companies was reported to have instructed its vessels not to stray farther up the Gulf than the Saudi terminal at Ras Tanura.[38] BP also attracted attention, when its tanker *British Renown*, taking part in a rescue operation[39] on the stricken Liberian-flag Very Large Crude Carrier (VLCC) *Tiburon*, was surprised by an Iranian aircraft near the Shah Allum shoal, a favourite Iranian hunting-ground,[40] and hit by missiles.

Of particular importance to worldwide navigation is the state of the world's insurance markets. The year 1984 was so active that the markets' response was both pronounced and rapid. War-risk premiums were repeatedly reported to be on the rise[41] for both hulls and cargoes. The GCC states were inevitably affected by this and generated one particularly interesting case. For the Kuwait Oil Tanker Company (KOTC), the news, at a time of generally increasing premiums, that Lloyd's had decided to include Kuwait in their definition of the War Zone[42] was especially unwelcome. KOTC concluded that on their fleet of 23 vessels, repairs were cheaper than insurance and promptly cancelled their London cover for the entire fleet.[43] In the case of cargo insurance, the GCC states as a group were particularly inventive, simply offering to replace outright[44] any cargo lost due to the war. This concept was soon invoked, when the *Gaz Fountain* was hit near Halul with a cargo of propane and butane.[45]

It was not merely insurance costs that were rising, but operating costs of all kinds, notably crew wages. The clamour begun earlier by the Japanese union had echoed around the world and seafarers' unions in many countries were pressing for material recognition for the dangerous, difficult conditions in the Gulf.[46] It was no longer, as one captain from a large fleet remarked to me, enough to give crews videos to watch and gymnasiums in which to work off steam while on board. In some cases, owners substituted specialist crews of 'mercenaries' (mostly Filipino, but sometimes German or British), ready to run almost any risk for the part of a voyage that lay in the war zone.[47] With freight costs tending sharply upwards, many crude-oil buyers began to shop around for discounts: despite formal OPEC condemnation of breaches of price discipline, such buyers were not always[48] disappointed.

Not all the incidents in Iranian waters involved tankers. The UK-flag bulk carrier *Charming* was hit by a missile while carrying a cargo of alumina in an Iranian-organized convoy going up to Bandar Khomeini. This port[49] was frequently a target for Iraqi attacks, if only because of the intermittently suspended Japanese-Iranian petrochemical plant which had been under construction there for some years.[50]

Another excitement in 1984 was the re-emergence of the subject of mines, this time not in the Gulf but across Arabia in the Red Sea. Despite conflicting statements from Tehran and press conjecture centring on Libya, it was never clear[51] who was responsible for laying them.

During 1985 there were 50 attacks on shipping as against 62 the previous year. After one attack at Kharg, the Iraq News Agency announced the terminal's destruction; this was widely[52] reported, although a later report confined itself to the installation's T-jetty on the eastern side of the island. There followed a series of raids on Kharg during the rest of August and September. It seems to have been the tenth raid which did the greatest damage, when the 'sea island' with its H-jetty was hit and a Korean tanker loading there set on fire. Shipping circles indicated that by the end of the month, despite the damage, operations were still possible, as in 1984[53] at the T-jetty where tankers loaded for shuttling crude to the makeshift terminals at Lavan and Sirri Islands down the Gulf. Both Lavan and Sirri were bombed by Iraq in 1986; Iraq's oil minister[54] claimed these operations had reduced Iran's exports to a level around 500–800,000 b/d, which may be compared[55] with the figure of some 3.1 million noted earlier. Iraq's Taha Yasin Ramadhan later expressed satisfaction[56] at having transformed Iran from an oil exporter into an importer.

Before the end of 1985 Iraq launched another attack[57] on the Nowruz offshore field near Kharg, first hit in 1983. At about this time, Iran introduced a procedure for subjecting shipping entering the Gulf to boarding and search, in the hope of intercepting military supplies intended directly or indirectly for Iraq. Tehran radio announced on 31 December that 'The Iranian navy has seized[58] a Danish ship in the Gulf of Oman which was carrying explosives destined for the ruling regime in Iraq and forced it to dock at one of the ports of the Islamic Republic of Iran.' A pattern emerged whereby shipping belonging to or destined for Kuwait was more or less assured of Iranian attention.[59] This explains Kuwait's move to involve the superpowers in its fortunes, first, by transferring some vessels like the *Bridgeton* to the US flag and, second, by chartering-in some Soviet vessels, in both cases anticipating some degree of protective surveillance.[60] When the US had offered escorts for Kuwaiti shipping much earlier, they were reportedly turned down.[61] By July 1987, however, both the Americans and European navies (including the Royal Navy's Armilla Patrol) were involved in active escort duties,[62] hunting for mines and similar activities. When British forces assembled in the Gulf in 1990, elements of the Armilla Patrol had been out there for most of the previous decade.[63]

Well before the events of 1987, changes had occurred in the way in which the struggle between tankers and their assailants was conducted. Crews and owners had been reflecting on how best to continue to operate in the Gulf. At the same time, both Iran and Iraq were developing new techniques. The Iranian air force had two particular flashes of inspiration. First, it began to send reconnaissance aircraft from Bandar Abbas, close to the Strait of Hormuz, down the Gulf of Oman, the approach route used by shipping making for the Gulf. Having spotted potential targets on their way towards Hormuz, they would alert F-4 strike aircraft back at their base. The second innovation was to look around for offshore oil facilities, island bases or production platforms on which the air force could station helicopters, equipped with missiles, sometimes improvised Maverick anti-tank models.[64] Their first success was with the Rostam field,[65] close to Qatar's offshore terminal at Halul. Of the 21 attacks attributed to Iran from January to May 1986, over half occurred in this area and many involved helicopters.

For its part, the Iraqi air force concentrated on finding ways of bringing the makeshift terminals at Sirri and Lavan within range of aircraft based near Basrah. In the course of 1986 they succeeded in reaching both, bombing Sirri and compelling Iran to redeploy its ULCC and VLCC storage tankers there.[66] It is not yet clear whether this feat was accomplished through inflight refuelling—there were reports that Iraq had received the Antonev fuel-carrier from the USSR—or through refuelling halfway down the Gulf, as the victims appeared to suspect.[67] The operation was repeated in November, when Larak—this time a 2,500 km round trip—was bombed for the first time and five tankers were hit.[68] Again there was widespread suspicion as to the techniques employed[69] and Tehran accused Riyadh of providing the Iraqis with fuel. On the day on which Iraq bombed Larak, a neutral oilfield—Abu Bukhoosh, offshore Abu Dhabi—was also bombed, with both Iran and Iraq denying responsibility: some evidence points to Iran[70] while other sources attribute the raid to Iraq.[71]

One large tanker fleet operator adopted a standard procedure whereby its craft never entered the Iranian Advice Area at all unless they had business there, in other words, cargo to load. In particular, they avoided the straight-line exit route to Hormuz from, say, Kuwait which inevitably ran through the Iranian zone. The preferred route lay along the coast of the Arabian peninsula, away from the deeper but vulnerable water on the Iranian side. This imposed an extra constraint on the larger vessels. A 350,000-tonne ULCC drawing 73 feet would have to beware of shoals and submerged ironmongery such as pipelines and wrecks on the Arabian side.[72]

Other rules while in the Gulf were, as far as possible, to move only at night, to observe rigid radio silence and to travel at maximum speed. On a voyage to Kuwait, perhaps to Ahmadi or Khafji, the drill was to pass the Strait of Hormuz in darkness, travelling straight across to Dubai, anchor for the following day and then set off in the evening to take up position off Das Island. The next day would be spent there, before departing at dusk for the Ras Tanura holding area to anchor there for the following day. Finally, that night the vessel would sail straight to its destination in Kuwait, arriving, all being well, at dawn.

Some shipowners proved particularly receptive to a suggestion that their craft employ a technique used during the Second World War by British aircraft to confuse enemy radar and later influenced by the Royal Navy's experience in the Falklands campaign: tankers were supplied with equipment to discharge clouds of aluminium chaff at the approach of an AS missile like Exocet, in order to divert it.[73]

Although most tanker owners and their crews were opposed to the idea of arming their ships, there appears to have been at least one exception: a Greek-flag cargo ship, the *Norassia Karsten*, was alleged to have fired a missile at a US helicopter in January 1987,[74] a charge the owners denied. The 1980s were in fact an extremely busy period for tanker owners—partly because of the boom in orthodox calls for shipping, partly because of the development of unusual kinds of employment.

In the mid-1980s the Saudi-supported concern Norbec took to chartering ULCCs and VLCCs in order to hold stocks of crude oil afloat and thus take advantage of any favourable movement in prices after cargoes were loaded. At its peak, the scheme involved some 35 vessels,[75] of a total tonnage probably approaching 10 million. It was phased out after Shaikh Yamani was replaced as oil minister in 1986.

Another important figure in the charter market was the National Iranian Tanker Company (NITC), which acquired vessels for shuttle service and crude storage at Iran's satellite terminals like Larak. Both VLCCs and ULCCs were used, including the ill-fated *Seawise Giant*, at the time she was hit[76] the largest ship afloat at 564,739 tonnes.

The freight charges vessels expect to earn are negotiated on the basis of a system called 'Worldscale', published twice a year in London and New York. The schedule tabulates rates for a list of frequently performed voyages such as Kharg/Rotterdam and Ras Tanura/Singapore. It is not particularly illuminating for the layman, as it does not specify the actual charge or reflect the real costs incurred in making the voyage. Nevertheless all the voyages are established on the basis of common assumptions and can easily be compared. They all bear the prefix 'W' (i.e. Worldscale) and in the context are easily recognized for what they are. Some examples appearing in the specialist press help to illustrate the use of Worldscale numbers:

(a) *Middle East Economic Survey* (MEES), indicating the terms on which both freight and insurance were being negotiated at a particular stage of the war, described the freight element as follows: 'Kharg to Europe: W70–80. This amounts to $2.15–$2.45/bbl, compared with W20, which yielded 60c/bbl.'

In another example from MEES, W45 was indicated as equalling $28,000/day.[77] *Lloyd's List* later cited a deal[78] whereby the *Pacific Valour* was to shift a cargo of 240,000 tonnes from the Gulf to Japan at W37, apparently a fairly durable rate, as the same publication reported W37–40 as the range over which discussion occurred[79] for a voyage from Kharg to Japan in 1984. In another report, Texaco fixed a ULCC of about 320,000 tonnes out of the Gulf at a rock-bottom rate of W20.[80] *Lloyd's List* had earlier reported the Kharg/northwest Europe voyage as particularly profitable: a VLCC at current rates could pick up $3.7 million at W40 which would yield a 25 per cent profit, compared with W27 a month earlier which would have earned just $1.5 million.[81]

(b) In the course of the Norbec hirings, one particularly low VLCC voyage rate was agreed from Yanbu' to Brazil at W23.5.

When the purpose is purely storage, the hire charge is computed on a daily basis, at so many thousand dollars per day. In May 1985 there was talk of NITC fixing another ULCC for Sirri storage at a rate slightly over $8,000/day.

(c) A description of the mechanics of 'netback' pricing (the device to conceal discounts which became widespread during the mid-1980s) in November 1985[82] gave the actual spot freight incurred on 1 tonne of Iranian Light crude travelling from Sirri to Rotterdam as $6.70. Reckoning seven barrels to the tonne, we can put the rough cost per bbl at just under $1.

To enter this eventful activity in 1986, the prospective owner would have had to find some $45 million to have a VLCC of 250,000 tonnes built new. By 1988 the figure would probably have been nearer $55 million.[83] Second-hand vessels sold for approximately one-third of this price and delivery was appreciably quicker.

For all the lucrative prospects, owners could not disregard the risks. From 1984 to 1988 there were 490 attacks registered,[84] of which 270 were attributed to Iraq and 220 to Iran. Not all were conducted by aircraft or the use of mines. There were a number of instances when Iran used 'gunboats' or other small craft, often mere motor launches. In one of these attacks, the victim was a Soviet merchantman;[85] in another, it was a vessel carrying sheep from Australia to Kuwait, the only recorded casualties being 300 sheep.[86] These craft were particularly active in the southern part of the Gulf, adding to the incentives to watchfulness in the Hormuz area. Iran also used the HY-2 Silkworm missile, supplied by China, reportedly positioning it so as to overlook the Strait. These missiles[87] were not used here but reappeared later near Fao when Iran fired at shipping at Kuwait's Mina al-Ahmadi.

Despite all the precautions taken by owners and crews, some vessels operating in or to the Gulf had exceptionally bad luck. The Kuwaiti-flag *Kazimah*,[88] the Turkish-flag *Atlas I*, South Korea's *Royal Colombo*, the Norwegian-Maltese *Torill* and the Iranian-flag *Dena*[89] were all attacked twice. So was Malta's *Harmony I*[90] while working on the Kharg-Sirri shuttle, a run on which Iraqi aircraft particularly concentrated.[91] While being towed back to Sirri after the first incident, *Harmony I* was attacked again and sank. One shuttle tanker, *Polikon*, was reported to have been hit for the fourth time and written off as a 'constructive total loss' (CTL).[92] When the Liberian shuttle tanker *Medusa* was hit on 10 June 1986, it was its third hit in nine months.[93] The insurance market identified shuttle craft as the most vulnerable in the Gulf[94] and raised premiums accordingly.[95]

The use of gunboats increased towards the end of the war. They were often crewed by Revolutionary Guards, generally employing small arms, for hit-and-run attacks, as described by *The Times* early in 1988:

> The pattern of attacks has not really changed at all; the Iraqis go for the Iranian oil tankers on the shuttle from Kharg Island to Larak Island, and the Iranians then retaliate with a burst of rocket-propelled grenades and machine gun fire against almost any vessel unaccompanied by a warship.[96]

The Western powers' recourse to warships yielded mixed results. The US Navy, despite the episode involving the *Stark* and another in which the USS *Vincennes*, believing herself to be under air attack, released the missile that shot down an Iran Air Airbus,[97] had the satisfaction of catching an Iranian makeshift mine-layer, the *Iran Ajr*, with its mines on board,[98] and confiscated both ship and cargo. It also succeeded, in retaliation for mine damage to the frigate *Samuel B. Roberts*, in demolishing two Iranian offshore oil platforms, Sirri and Sassan.[99] The effect of these operations must surely have been to shorten the war.

8
The War of the Cities

As if the casualties being inflicted on members of the armed forces of both parties, not to speak of the crews of the largely neutral shipping involved, were not enough to satisfy the appetite of some mythical god of war, further carnage was developing, mainly in the latter half of the war, among the civilian populations of both countries. First the opposing air forces were brought into play and then the two sides' arsenals of short- and medium-range projectiles, conventional artillery and surface/surface missiles. The use of airpower by both belligerents, in addition to supporting their ground forces, was largely designed, first, to damage the enemy's morale and, second, in retaliation for acts perpetrated by the other side.

Iraq's attacks on shipping and ports fit into both categories. Denial of oil exports to Iran was conceived both as a means of rendering that country harder to govern, thus denting the morale of its leadership, and as retaliation for the severance of Iraq's own crude export systems by Iran.

As we have seen, Iran's assault on the enemy's oil-export apparatus was conducted in three ways: by diplomatic action, through Syria, which Iran persuaded to close the pipeline that led to Homs, branching there to Banias and Tripoli; by naval action in the Shatt al-Arab; and by aerial attacks on the refinery and crude-oil treating plant at Kirkuk in 1980.[1] Here again Iran's strategic aim had been to render its opponent harder to govern. Part of this sequence of attacks was a form of retaliation, this time for what Iran perceived as the provocative act of the initial invasion. The attack on Kirkuk—a straightforward economic target—was accompanied by others on Baghdad itself, where the Daurah refinery was hit in September 1980.

Iraq was for some years generally regarded as possessing air superiority, although this was not translated into any particular battlefield advantage. Not only did Iraq,[2] by the middle of the war, enjoy a six-to-one advantage in conventional aircraft, bombers (the Tupolev Tu-22s and Tu-16s, the Ilyushin Il-28s) and fighter-bombers (MiG-23s, Su-7s and Su-20s); it was also reported to have an assortment of Soviet-made missiles.

Apart from the AS-4 Kitchen and the AS-6 Kingfish, both suitable for use with the Tu-22 and Tu-16, the Iraqis also had a number of surface/surface systems, the Scud-B, the Frog-7 and the SS-12. There was widespread interest in the possibility that weapons in the last group might be used to attack Iranian population or economic centres. For the most part, however, these systems were of limited range, designed for the battlefield rather than for strategic use and more suitable for targets in Khuzestan than Tehran. Indeed, the Iranian capital was in the main attacked by aircraft, at least early in the war. There were reports that Iraq had received another Russian system, the SS-21, though specialist circles are uncertain about this and arrival was not independently confirmed—perhaps the confusion arose from the last digits of the SS-12 having been inverted.

Turning from missiles (or, as the specialist world has it, delivery systems) to warheads, the SS-type device can be made to carry either a conventional (high-explosive of some sort) or a nuclear warhead. This subject bred its share of Middle Eastern coffee-shop rumours, as for example in April 1984, when a London newspaper—citing unidentified Gulf sources—reported that Iran was making a nuclear bomb, to be available two years later.[3] There were also unconfirmed reports that West Germany's KWU was studying a resumption of work on the two reactors ordered by the Shah's government in 1976 and suspended after the revolution.

From the very beginning of the war in 1980, both countries' air forces were active over one another's capitals. Reports from Tehran on 22 September told of an attack that day on the city's Mehrabad airport, which, like a number of others around the world, was used partly for civil and partly for military purposes. The following day, Iranian aircraft were over Baghdad. Official Iraqi statements reported them as having bombed residential areas, a charge denied by Tehran. Although air-raid sirens were often heard in Baghdad during the months that followed, both sides' air forces tended initially to focus on economic targets such as the oil installations in northern Iraq and the Abadan refinery.

Attacks on some of Iran's smaller towns and cities occurred comparatively early in the war, but apart from the first two days, it is not clear when either side decided to strike at population centres as a strategic option. The main instruments available to Iran were aircraft and artillery. In the light of Iraq's generally accepted air superiority, it is hardly surprising that Iran had frequent recourse to its artillery, particularly at Basrah, the one large population centre within easy range. At some stage, however, Iran secured supplies of what were believed to be Soviet-originating missiles, possibly from Syria, to which the USSR had delivered such weapons. Another possible channel for the supply of Soviet equipment was Libya.[4] One system specifically indicated was the Scud-B, which was at one stage used by both sides for attacks on each other's capitals.

Basrah had a hard war, first by being caught up on the fringes of routine military engagements and later by itself becoming a direct target for bombardment. Whether this was initially due to the economic activity centred on it, such as shipping, or an Iranian desire to intimidate the inhabitants was never clear.

Nor is it clear when Iraq began to employ surface/surface missiles to attack population centres. By 1983 this appears to have been happening, with reports[5] of the use of such weapons at both Dezful and Gilan-e-Gharb. Later that year reports began to circulate in Baghdad of the coming availability of the SS-21.

It was soon evident that the missile option was not the only one Iraq was considering. Iraqi aircraft overflew Tehran and the holy city of Qom in January 1984 to demonstrate to the Iranian leadership that no part of Iran, except remote towns like Mashhad,[6] was beyond their reach. The following month, after an appeal by Iranian opposition leader Rajavi, Iraq announced that for seven days it would halt attacks on towns. This seems not to have impressed the Iranians, who the following day attacked Baaquba, near Baghdad. In June, after a bloody episode in which both Basrah and nearby Khorramshahr received ferocious hammerings, the

United Nations mediated a joint undertaking to abstain from civilian targets.

Nevertheless the following year attacks continued. Shortly after Iraq had bombed a steel plant near Ahwaz and the nuclear construction site at Bushehr, Iran shelled Basrah. Tit-for-tat actions followed, including a missile attack on Baghdad. In June President Saddam announced proposals for a further and unilateral two-week halt to attacks on cities, hoping Iran might be tempted to discuss an overall settlement of the conflict, complete with a cease-fire, the withdrawal of both armies, an exchange of prisoners, negotiations towards a peace treaty, the resumption of normal relations and non-interference in one another's affairs. Iran had no intention of being drawn into peace talks unless its preconditions were accepted. These were made clear to all would-be mediators and featured, in particular, the payment of several billion dollars as reparations and the 'punishment of the aggressor', understood to mean the removal from office of both Saddam himself and the entire Baathist regime through some form of Nuremberg mechanism.[7] Needless to say, the initiative led nowhere and the fighting continued.

The accounts of these operations issued for home consumption by the two sides were regularly monitored by the BBC. Fragments from the *Summary* of 20 March 1986 give an idea of the flavour:

> *Baghdad, 17 March*
> After our daring air force aircraft launched a series of attacks to spread death and terror among the Khomeyniites, squadrons of our aircraft raided a concentration of enemy machinery in an area south of Ahwaz at 09.40 today. Our helicopter gunships flew 118 missions and returned safely to base. The forces of tyranny and aggression shelled the Mandali suburbs, Khurmal and Darband-e-Khan with long-range artillery.
>
> *Tehran, 17 March*
> [Iraqi target, citing map references] . . . was bombed by several fighter bombers of the Air Force of the Islamic Republic of Iran. The criminal Baathists of Iraq sustained heavy and irreparable damage. The selfless pilots of Islam, on their return from their victorious mission, hit and shot down two enemy helicopters in the region. Our aircraft all returned safely to their bases.

Apart from the aircraft involved in military operations, the civil aviation community now had to serve the major centres—if it could—by new rules. Mehrabad, as previously mentioned, was attacked at the very beginning of the war. At Baghdad's international airport, the national carrier, Iraqi Airways, concerned lest there be raids by Iran in retaliation, took steps to disperse its fleet. Some of the larger aircraft were transferred to the parking area at Amman airport, for a long time the only point from which Iraq could be entered from abroad, as foreign airlines were hesitant about using traditional air corridors which ran close to a war zone. As the war proceeded, however, several started operating by night and with all the lights blacked out. Occasionally an airline would suspend its operations to one or both capitals, but most found the volume of traffic, even at depressed war-time levels, justified their continuing to fly.

There was obviously a danger that domestic flights in the two countries might run foul of incursions by enemy aircraft. As Iraqi Airways suspended flights to Mosul and Basrah early in the war, this was not a major hazard in Iraq. In 1986, however, a Fokker F-27 belonging to the Iranian domestic operator Asseman and carrying a party of Iranian deputies and officials on a visit to the battlefront in the south was shot down and 40 people[8] reported killed.

While the war was on, life in the major Iraqi centres was austere, with reminders of the battlefront seldom far away. The streets and all government offices were crowded with people in uniform; there were frequent air-raid sirens and night-time blackouts. The ubiquitous billboard portraits of President Saddam often showed him in a green uniform. Military vehicles were no rarity in the streets, while press and television maintained a steady flow of information on how things were going on the different fronts.

Many necessities of life were hard to find or more expensive than before the war; even motor fuels, derived from the one substance the country possessed in abundance, rose in price as the war pursued its course.[9] In Basrah,[10] Baghdad and wherever Shiite communities were to be found, the streets were festooned with black banners lettered in white, announcing the name of a new 'martyr' and at which front he had fallen. Around airfields and other installations of national importance emplacements of the distinctive Soviet ZSU-23 multi-barrelled anti-aircraft guns could be seen.

Baghdad underwent a major physical transformation early in the war, partly in preparation for the summit conference of the Non-Aligned Movement (NAM), programmed for September 1982.[11] In the end,[12] the danger of Iranian air raids meant that the conference was postponed until the following year and transferred to New Delhi.

Air raids were not the only hazard to which central Baghdad was exposed. In June the facade was blown out of the Iraqi Planning Ministry building[13] by what seems to have been a car bomb, although it is not clear whether this was the result of direct Iranian action or that of some domestic opposition group. Both Damascus and Tehran provided encouragement for anti-Baghdad Iraqi groups. In Syria, the establishment of an emigré provisional government led by Brigadier-General Naqib was reported in 1982[14] and two years later an Iraqi National Democratic Front, including such elements as the KDP,[15] was set up there.

Meanwhile, Tehran was equally active, fostering the Supreme Assembly of the Islamic Revolution in Iraq (SAIRI), under exiled Iraqi Shiite leader Hojjatoleslam Baqer al-Hakim.[16] (Some years later, in 1988, Hojjatoleslam al-Hakim's brother, Mahdi, also a prominent dissident, was mysteriously assassinated in Khartoum.[17] This maintained a pattern of vulnerability for the family begun with the execution in June 1983 of five clergymen members and the arrest of many of their relations at about the same time.) Tehran also broadcast calls for Iraqis to ambush Iraqi troops and seize their arms, to rob the rich and put pressure on resident foreigners[18] to force them to leave.

Any resulting sense of insecurity among the foreign communities may have been equalled in some of the upper reaches of the Iraqi establishment. In 1982 eight ministers lost their jobs in a mid-year cabinet reshuffle.[19] Later on, the former industry and interior ministers, Hassan Ali and Saadun Shaker, were relieved of their posts, effective from 4 August 1987.

In 1988 Lieutenant-General Maher Abd al-Rashid, war hero of campaigns in the south[20] and described at one time as having fought a 'perfect tank battle' in the desert near the Howeizah marshes, was said to be under house arrest.[21] Another unexpected disappearance from the scene was that of the defence minister himself, Adnan Khairallah, the President's cousin, reported early in 1989 as having been dismissed[22] and later to have died in a helicopter crash. By 5 May 1989, according to *The Times*, Khairallah had been replaced as defence minister by General Shanshal. Even President Saddam was reported in May 1987 to have survived an assassination attempt.[23]

The year 1986 saw bomb incidents and casualities in Iran:[24] in Qom by a shrine and in Tehran at Ferdowsi Square. Life for Iranians was austere.

Gasoline had been rationed since 1978, when oil output was reported so low that domestic needs could not be met.[25] Power cuts were not uncommon, following the bombing of power stations.[26] Basic foodstuffs were in short supply and the local vodka and wine (not to mention imported liquor) that used to be readily available before the revolution in Tehran and a number of other centres were prohibited and could be found only on the black market.[27] The Imam's countenance in Iran mirrored that of Saddam in Iraq.

Gradually, and despite the readiness with which the ranks of the *baseej* had been filled from the towns and villages, a sense of war-weariness was apparent in Iran; in 1984 there were even newspaper reports of a loss of the will to fight in the ranks.[28] There were certainly those who felt the time had come to end the war, a courageous stance in view of the Imam's known determination to keep going. One who came out into the open after having lost his political way somewhat was former premier Bazargan.[29]

Another factor which depressed general morale was the occasional rumble of disagreement[30] at the top of the regime, notably involving President Khamenei and the formidable Majles Speaker Rafsanjani.[31] To some extent, it was a repetition of the atmosphere early in the war when the first president, Bani Sadr, was at odds with many other senior figures and the clergy in general.[32] Much of this new atmosphere, whose existence was officially denied,[33] was generated by the issue of the succession to the long-ailing Khomeini. The designated successor, Ayatollah Montazeri, gradually ceased to command the Imam's confidence and, wearying of it all,[34] announced his withdrawal early in 1989. Early in June the Imam died[35] and the new dominant figure in Iran was Hashemi Rafsanjani, now the Islamic Republic's president.

Conditions within Iran during the war years were not free from criticism.[36] Reports by the United Nations and Amnesty International detailed human rights abuses under the regime, one citing 14,000 cases of execution or torture over a six-year period. Amnesty International had harsh words for Iraq also.[37] The fate awaiting ordinary civilians in both countries during the 1980s was decidedly uncertain.

Map IV. Important Pipelines, 1980

9
The Oil Industry in the 1980s

Iraq Plugs the Gaps

For the first years or so of the war, oil and gas operations, setting aside the damaged Gulf terminals and the dislocation of the northern pipeline, continued very much as during the previous decade. In the 1970s, however, the setting up of contingency provisions for export in the event of difficulties with conventional arrangements was not uncommon. In addition to Saudi Arabia's installation of the east-west 'Petroline' linking the huge Ghawar oilfield to the Red Sea coast, Iraq had also laid a strategic line to enable oil from its southern fields to be sent out via Syria in the event of difficulties in the Gulf.[1] The difficulties that now materialized were, however, of another kind. On 10 April 1982, after Basrah crude had been regularly loading at Tripoli (Lebanon) and Banias for some while, the Syrian government—in a gesture of solidarity with Iran—closed the line and with it both ports.[2] This confined Iraq to the one remaining exit route, via Turkey to Ceyhan, and in effect to what could be delivered from Kirkuk and neighbouring fields to the K1 pumping station at which the northern lines both began.

For several years, Iraq simply put up with the situation. Then in 1986 there were two interesting developments. The first related to the damaged Iraqi terminals. Iran gave the impression that it believed Iraq was planning to replace them with single-buoy moorings (SBMs), a subject on which rumour had been rife[3] since 1980. Reports now came in that the wreckage of the terminals had been bombed.[4] The second development was that some Basrah crude finally reached the open sea. Although there were plans to expand the capacity of the Ceyhan line to a total of 1.5 million b/d, this offered no solution as the system was physically dedicated to northern crudes which, when expansion work was complete, it continued to handle. In addition, the strategic-line concept presupposed the availability of the Syrian line with which it linked up at the K3 pump station. This presented a sizeable engineering problem: although INOC managed to find a solution, it was really little more than a stopgap and the long-term answer proved to be the negotiation of an arrangement with Saudi Arabia. In addition to the other help the Saudis were providing to Iraq in different forms, including loans of oil, they allowed Iraq to construct a 500,000 b/d spur line to link up with Petroline and thence the Red Sea. This was in due course designated Iraq Pipeline Saudi Arabia No 1 (IPSA-1)[5] as distinct from a later scheme whereby a separate and entirely Iraqi line to handle 1.6 million b/d duplicating Petroline was to be built and called IPSA-2.[6] Basrah Light crude became available to Iraq's customers again from 28–29 September 1985.[7] The price at Yanbu' was \$17.30 per bbl, while at Ceyhan it was \$17.50 as against the Turkish line's normal feed, Kirkuk, at \$17.60.[8] By achieving availability once again in the Mediterranean, Iraq hoped to be able to raise its output to 2 million b/d for 1986, though this figure fell well short of its potential, widely estimated at nearer 4 million b/d.[9]

Other schemes were contemplated in the early 1980s. Some crude was exported by truck, to Turkey[10] and Jugoslavia, and also to Aqaba in Jordan for onward shipment to Brazil and elsewhere.[11] Aqaba, close to Israeli territory, was also the focus of a more orthodox but abortive export scheme, involving a pipeline across the desert into Jordan and complicated negotiations whereby the US engineered Israeli undertakings not to attempt to sabotage the line despite a well-known Israeli antipathy for Iraq.[12] One possibility might then have been a link-up with the Sumed pipeline across Egypt from the other side of the Red Sea to Alexandria, but the evidence for this is confined to a report in MEES on 25 March 1985 and the Aqaba project was never implemented anyway.

One project Iraq was no doubt relieved to see never brought to pass concerned the much spoken-of Majnun field. Iran, having captured the tract of marshland under which this major accumulation lay,[13] was reported by its oil minister, Muhammad Gharazi, as contemplating bringing it into production.[14] However, contemplation appears to be as far as the idea ever got.

Iraqi oil-processing plant also came under fire from Iran. The Basrah refinery was put out of action, although that at Daurah near Baghdad seems to have escaped damage.[15] The production centre at Kirkuk was attacked early on and thereafter ringed with air defences including—according to one foreign visitor from a company providing contract services to INOC—barrage balloons. The gas/oil separation plant and K1 pump station sustained major damage and still bore the scars some two years after the attack.

Iran Also Seeks Access to the Open Sea

Iraq did not manage to immobilize Iran's oil-production apparatus on a scale similar to the damage it had itself sustained in 1980. It was not until well into the war that serious attacks were first launched on the principal Iranian terminal at Kharg Island, a nut that proved particularly difficult to crack. The new terminals down the Gulf were easier targets and NIOC was obliged to attempt the contingency planning of some onshore schemes to ensure the continuing arrival of Khuzestan crudes at the Gulf coast. In a bid to bypass the Strait of Hormuz altogether, a 1 million b/d pipeline was proposed, running from the Gach Saran field via the Gurreh pump station nearby along to Jask, out on the Gulf of Oman.[16] More modest were the proposals to install SBMs off Ganaveh, the onshore starting-point for the pipeline across to Kharg[17] and other short-range schemes to reach Bandar Asaluyeh[18] and Bandar Taheri.[19] Another idea was to adapt a gas line laid earlier to the USSR, so as to carry crude instead.[20] There were also plans, as in Iraq's case, to build a line through Turkey to the Mediterranean, but this idea ran into financial trouble.[21]

After the battering the Abadan refinery received in 1980, Iran's other refineries were attacked in the course of Iraq's periodic operations against 'economic targets',[22] in an effort to create shortages within the country and accentuate the need for fuel rationing.[23] It was only in 1989 that anything was heard publicly of schemes to repair the damage, though anyone familiar with the gossip in contracting circles would not have been surprised to learn of NIOC's plans to tackle the subject: first, to re-operationalize Abadan itself—apparently entirely with home-grown resources and without recourse to imported knowhow—where the damage was assessed at $9 billion,[24] and, second, to add new refineries at Arak and Bandar Abbas.

Iranian crude production during the war never returned to anything near the 6 million b/d level, and although the level fluctuated throughout the 1980s, the original OPEC production quota assigned to Iran in 1982 of 1.2 million proved a reasonably accurate estimate. By late August 1982, with the start of Iraqi attacks on Kharg Island (which in Iran's peak production years in the 1970s handled nearly all the country's export volume), the volume was down to a meagre 700,000 b/d[25] from 1.8 million the preceding month. Nevertheless at one stage in the build-up to the OPEC conference of December 1987, Oil Minister Aghazadeh, pressing for a price rise, threatened to flood the market, saying Iran 'could produce 4 million barrels a day'.[26]

Iraq was undoubtedly successful in causing shortages of refined products in Iran. At different times during the war, reports came in of the need for jet fuel.[27] One solution adopted by Iran appears to have been to send crude off to Aden or other refining centres abroad like Singapore and Italy for processing into products to be sent back to Iran[28] and discharged at storage facilities available at terminals on Gulf islands like Lavan.

OPEC

The 1980s were a decade of trial for OPEC. The organization had started well enough, projected from the springboard of the prosperous 1970s, although the 1979–80 price explosion had an unexpectedly severe impact on oil demand,[29] which had been looking particularly promising as 1979 was drawing to a close. This and the fact that there was some disarray within OPEC over prices made 1981 a difficult year.

Saudi Arabia was going flat out at 10 million b/d,[30] to the dismay of several members who would have liked this reduced to ease the market glut they felt the Saudis had created. It can be argued (and was so argued[31] by the Saudis) that any over-supply had been provoked by the encouragement that the over-rapid price increase in the 1970s had given both to fuel conservation on the part of consumers and to the economics of projects to develop substitutes for oil and especially OPEC oil. Saudi

Arabia and other members also disagreed over the interpretation of the 'marker' crude price set at the previous year's OPEC conference in Bali, with a resolution so widely phrased that each member could do whatever it chose.[32] At the Geneva conference in October, the members split the difference of $4,[33] settling on a new reunified marker level of $34.00, which for all concerned, producer and consumer alike, was an improvement on the previous uncertainty.

In March the following year came the inauguration at the Vienna conference of a new OPEC institution. This was the production ceiling, a limit within which the entire organization was to contrive to hold its aggregated output, with every member adhering to its own assigned individual quota. Set in the first instance at 17.5 million b/d, the ceiling remained at about this level, subject to variations (sometimes just for a specific quarter), for most of the decade.[34]

As implied above, another effect of generally high crude prices earlier on had been to encourage the economics of developing oil accumulations discovered outside the OPEC countries—not only in the industrialized world, in places like the North Sea, but also in some countries close to OPEC members, like Oman and Mexico. The advance of these oil-producing countries, which became known collectively as the 'non-OPEC' group, was a notable feature of the early 1980s. By the December 1985 OPEC conference in Geneva, OPEC members were so exasperated by the inroads made in their traditional market by the 'non-OPEC' group that they evolved an entirely new, if short-lived, strategy. This was henceforth to seek some given minimum share of the market represented by demand outside the CPEs, instead of the situation hitherto, whereby OPEC acted as the 'residual supplier', providing what was still needed after non-OPEC suppliers had eaten their fill. Before this decision[35] was reached, Saudi Arabia had reacted to market pressures[36] by initiating a number of netback deals, amounting to discounts on official prices, with the aim of restoring its sagging output to the current official quota level of 4.35 million b/d.[37]

This objective was indeed achieved, but with it came a spectacular collapse in crude prices,[38] which lasted from 1985 into the following year, when there were calls from the Gulf states for Britain to cut its North Sea output[39] 'as part of a global effort between OPEC and non-OPEC exporters to restabilize the world oil market'. Dr Alirio Parra, a senior Venezuelan technocrat, described the atmosphere as one in which 'the free play of market forces took over as OPEC producers adopted a short-term strategy to regain market share'.[40] One lesson he identified from this experience was that 'the demand response to prices is relatively inelastic in the short term'.

While at the end of 1985 there were fears that the price of crude would drop as far as $15, by the end of July 1986 a leading article in *The Times*[41] reported that Brent crude stood at 'a paltry nine dollars a barrel'. A week later, responding to an OPEC decision to cut production, the Brent price was back up to $15.

Effective though it proved, the device of holding down output in the hope of driving up prices brought with it problems not to be found in any economics textbook: the global output ceiling, if set realistically in relation to demand, can easily prove too small to be divided amicably. From the outset, Nigeria and Iran sought quota improvements and in the late 1980s there arose some aggrieved discussion of the UAE's insistence on setting its own unilateral quota (1.5 million as against the 0.948 set by the Vienna conference in 1987)[42] and vehement debate with both Iraq and Iran, neither of which was willing to accept a lower quota than the other's.[43] The year 1988 was another difficult one for OPEC, not only due to problems over quotas but also because of signs of a possible sequel to 1986. Prices again began to weaken. OPEC Secretary-General Subroto issued a warning to the effect that over-production 'by various OPEC countries including the UAE' had raised the possibility that 'countries with big capacity will also increase their production' in an attempt to restore discipline.[44] Saudi Arabia, as its new oil minister Hisham Nazer stressed, had no intention of acting—as it once had—as OPEC's 'swing producer', expanding production if the aggregate OPEC output fell short of the ceiling and cutting back if the ceiling seemed likely to be exceeded.[45]

One difficulty in establishing a quota for Iraq was the breakdown of its export apparatus in 1980–82. Thus it was that in 1983, accepting a quota of 1.2 million against Iran's 2.4, Iraq entered a caveat that its figure should be revised upwards if and when the Syrian line reopened.[46] One strange chapter in this story came to be written at Geneva in 1986 when a difficult problem of ceiling division in conference was overcome by Iran's proposal that Iraq be freed from the discipline of a quota, enabling it to produce as much as it chose. Given the atmosphere between the two countries, if only over quotas, the initiative took both delegates and journalists by surprise. One commented on 'Iran's astonishing gesture in accepting . . . while all the other OPEC members bound themselves to return to the quotas of 1984.'[47]

An alternative analysis saw this as the culmination of a highly effective episode of Iranian diplomacy, aimed at isolating Iraq within OPEC. The GCC producers in OPEC were also progressively isolated as Iran tried to convince non-Gulf members that the endorsement of Saudi low price policies was no longer in their interest.

The Consuming World

The United States is not only the world's largest oil-consuming country[48] (15.6 million b/d in 1980) but also the largest oil importer (5.4 million b/d in 1980) despite its substantial production (second only to the USSR). The other notable consuming regions are Western Europe (12.9 million b/d in 1980) and Japan (4.5 million b/d in 1980). Japan has no significant domestic oil resources, which explains its wartime bid to seize those of the Dutch East Indies in 1942. It has recently sought to overcome this deficiency by obtaining equity access to Middle Eastern resources, as in Kuwait and Abu Dhabi. Western Europe (whose main petroleum resources are confined to the North Sea and a handful of small onshore fields) has some other limited energy options, but, like Japan, is mainly dependent on imported oil.

In 1976 demand outside the CPEs peaked at just over 52 million b/d,[49] falling to some 45 million in 1982–83. As the end of the decade approached, natural growth was carrying this towards a round figure of 50 million. From this had to be deducted another element in a piece of crude arithmetic which came to dominate OPEC's expectations. This factor was the approximately 30 million b/d which represented the supply of crude from non-OPEC sources, leaving a balance—the quantity to be provided from other sources, meaning OPEC or the drawdown of stocks, which were sometimes very substantial.[50] Analyses on these lines appeared periodically in the IEA's *Oil Market Report* (OMR).[51] Conversely there were occasions when world oil production reached levels which could not be explained by known consumption forecasts, as IEA member governments established or topped up strategic stocks against the uncertainties of life, like the US government's vast Strategic Petroleum Reserve (SPR). By 1987 the OMR September analysis showed a forecast for the full year as follows (all figures in million b/d):[52]

Demand outside CPEs	47.3
Non-OPEC supply	28.4
Required from other sources	18.9

This was against a background in which the expected OPEC output for the third quarter was 19 million b/d (against an OPEC global ceiling of 16.6) and the stockbuild forecast was over 2 million.

To take one of the world's most important individual consuming countries, during the first half of the 1980s Japan's imports of crude were maintained in the region of 4 million b/d, running on from 4.83 million b/d in 1979.[53] The largest single source was Saudi Arabia, followed by the UAE. As Iran and Iraq were further sources, Japan sought to maintain strong bilateral ties with both countries. In Iraq, Mitsubishi was active not only in petroleum dealings but in construction contracting, while in Iran, Mitsui's interest in crude oil acquisition was complemented by involvement in the joint Iranian-Japanese petrochemical scheme at Bandar Khomeini.[54]

As far as the oil itself was concerned, there was a whiff of competition in the air, with Iran reported early in 1984 as proposing a counter-purchase system for buying Japanese goods and services as an incentive for Japanese buyers to renew contracts for Iranian crude.[55] This theme was being maintained a year later, by which time it was clear that the incentive that held the greatest attraction for Japanese buyers was still the discount, particularly as the hazards of lifting oil in the Gulf were making insurance so expensive.[56] An indication of the order of discount the Iranians were willing to accord comes from a report of a deal with Mitsui for two cargoes of Iranian Heavy crude.[57] The delivered price was described as being on a par with the official f.o.b. price, in other words, the Iranians were discounting by the entire freight element. This means a figure somewhat below $2.00/bbl, which may be compared with discounts available the previous year for Iranian Heavy of 80–85 cents/bbl. On a different plane, it may be noted that just before 1984's seven-country G7 economic summit in London, Japan was clamouring for access, in the event of a supply disruption, to the US SPR.[58] The commercial arena saw one unusual barter deal in 1986, in which a Japanese firm, Tomen, featured as middleman in an Iran-New Zealand exchange of oil for meat.[59]

Currency Problems

As we have already seen, the US dollar went through a bad patch in 1986, in company with the oil market. In this year the US economy experienced serious trade and current-account deficits: the $170 billion trade deficit for the first nine months was due in particular to an increase in non-oil imports. This was sufficiently pronounced to offset the drop in the oil import bill following the recent price collapse.[60] By the third quarter, the dollar stood dramatically low against the German mark and the Japanese yen.

By late 1987 the situation was again serious, this time attributed[61] to the spectacular stock-market crash which echoed around the world. Comment on the dollar had been sombre for much of the year in the train of developments the previous autumn. Iran's Prime Minister Moussavi[62] was pressing for the dollar to be discarded as the unit of account for pricing oil, in favour of indexation to a basket of stronger currencies. Iran found that recent improvements in the price of oil had been negated by the fall in the value of the dollar. It now urged immediate action to bring about a price increase by abandoning the production ceiling increases programmed for the rest of that year. As the 1987 Vienna OPEC conference approached, there was further talk of a price increase, although the mainstream preference was for waiting until demand stood at a level capable of sustaining one.

10
The War of Words

The exertions of the opposing armies, air forces and other formations did not go unmatched by those of their supporting information media. Both sides made the fullest possible use of radio and television and any other means available to reinforce the spirit of resistance of their respective populations, and at the same time to depress that of the opposition.

Apart from the contributions they both made to raising the general temperature along the international border in the months leading up to the Iraqi onslaught in September 1980, the first blow appears to have been struck in the same month by Baghdad with a broadcast announcement of the death of Ayatollah Khomeini.[1] The Iraqis had previously given a trial outing to a line of attack which underwent persistent repetition as the war proceeded. On 19 September the Baghdad daily newspaper *al-Jumhuriyah* came out with a front-page reference to the enemy as *al-Unsuriyin al-Furs* (the racist Persians), a theme which came the more readily to the lips of radio and television announcers for the uncongenial rasp to which the phonetics of the expression lend themselves.

One favourite medium of attack in Iraq was the newspaper cartoon, with spectacularly uncomplimentary portrayals in the Baath Party newspaper *al-Thawrah* of the three chief objects of the leadership's hostility—Israeli Prime Minister Menachem Begin, Syria's President Hafez al-Asad and of course Khomeini, on one occasion presented, at a time when Iran was widely reported to be cutting oil prices to boost sales volume, as a door-to-door kerosine street pedlar calling out, '*Naft! Bilbalaash!*' (Oil! Free!). Later came Tehran's turn to have a tilt at Baghdad, this time using the radio in a commentary on the debate within OPEC in embattled 1986: 'In addition the Baathist traitors in Iraq demanded a larger oil-exporting quota through Saudi pipelines . . . [and] . . . openly announced that they would not adhere to OPEC resolutions.'[2] Occasionally Tehran varied the epithets a little. A week or two earlier, the Baghdad regime had been called 'Baathist-Zionist',[3] a compliment which was sometimes returned by Baghdad, as when it described the Iranians as 'the servants of Zionism',[4] a change from such images as 'the tyrant gang in Tehran'[5] and 'the spiteful charlatans'.[6] Baghdad's accusing Tehran of being the servants of Zionism was a reference to the Israeli role in resupplying Iran's armed forces in what came to be known as Irangate; in a broadcast in February 1986, Baghdad radio named the Israeli agent Colonel Nimrudi,[7] who featured prominently in separate coverage of the affair.[8] At one stage early on in the war, Tehran, amid the general abuse, urged the Iraqi population to withhold payment of taxes and electricity bills, and to rise up against the government.[9]

The Tehran media were not the only source of hostility to Saddam. In 1988 the Iranian regime's London-based critic, the right-wing Persian-language *Kayhan*, attacked Iraq's use of chemical weapons, on one occasion printing a cartoon featuring the president's face juxtaposed with a skull.

Both sides would broadcast interviews with senior personalities, in the course of which they gave interpretations of current developments. BBC monitor reports early in 1986 provided examples of this. In January President Saddam, speaking on Iraq's Army Day, referred to having reminded 'the gang . . . in Iran' of 'the disasters it sustained in the July 1984 battles'. In March it was the turn of Iranian Foreign Minister Velayati, who described reports that Syria was to reopen Iraq's westbound pipeline as 'lies'.

Both sides did their best through the medium of television to ensure that the general population's familiarity with the war was not confined to news of casualties. Throughout Iraq and even in neighbouring Kuwait, it was possible to share something of the flavour of the front line through the early evening transmissions of *Suwar min al-maarakah* (pictures from the battle), which followed the news and which, in addition to depicting the bombardment of distant Iranian positions by Iraqi artillery, contributed some human interest with groups of soldiers, sometimes regulars, sometimes from the People's Army, having a brew-up in some operational area, waving to the camera and generally putting on a show.

One story that surfaced late in 1982 told of a bold Iraqi initiative at the time. This received only limited press attention outside the country despite the novelty. What Baghdad now proposed was that both sides hold referendums to decide the legitimacy of their respective regimes. A London daily reported that 4 million Baghdadis—nearly one-third of the country's population—poured into the streets in what the media claimed was a solid show of support for Saddam. Iranian premier Moussavi's response was to say that Iran was only too ready to help make the referendum possible.

Irony was not frequently encountered in the literature on the Iran-Iraq conflict. One interesting case nevertheless arose on Tehran radio, in an excerpt from Communiqué No 47 of 25 September 1980. It ran:

> We hereby inform the valiant and heroic nation of Iran that last night 50 invading Iraqi tanks which, by the will of God, had become bogged down in mud were attacked by the fighting Iranian forces; all of them were destroyed and the invaders fled. The bogging down of the tanks in the mud reminds one of the immobilization of the helicopters of the master of Saddam in Tabas.[10]

11
Mediation Attempts

Throughout the war, there were constant efforts to limit the devastation it was causing or to bring it to a halt altogether. At one stage, Iraq was seen by foreign observers as seeking to bring Iran to the negotiating table by force.[1]

Among the weightiest of the many parties to embark upon the formidable task of attempting to bring the slaughter to a halt, by mediating between the belligerents, were the UN, the Non-Aligned Movement (NAM), the GCC and the Islamic Conference Organization (ICO).

In the UN, the inevitable instrument of most attempts at mediation was the Security Council resolution, in which two themes were intertwined: ending the war and freedom of navigation. In one case, a further strain intruded: the attribution of guilt. The GCC states—in the diplomatic as in other fields, generally inclined to support Iraq—had lobbied vigorously during the run-up to the Security Council vote[2] for the emergence of a form of words incorporating censure of Iran. In the event,[3] as we saw earlier, this did not appear; there was merely condemnation of attacks on shipping moving to and from Saudi Arabia and Kuwait. Neither was there any reference to Iraq, and it seems from the wording of one report on the proceedings,[4] to the effect that 'the UN implicitly criticizes Iran', that here was a resolution with human characteristics, one capable of being all things to all men.

The previous year (1983) another Security Council resolution[5] had been attended by similar backstage activity.[6] Then a visit to the area by a UN mission had to be postponed, as Iran expressed its unwillingness to discuss the political issues related to the war.[7]

The war presented the UN with other occasions for activity, quite apart from the content of this and other resolutions. First, there was the 'war of the cities', in which the UN's attention was attracted by reports of the considerable civilian casualties on both sides. As mentioned previously, Iran's share in this was mainly through the shelling of Basrah and other towns, while Iraq was conspicuous in the use of its abundant stocks of missiles and the decided air superiority afforded by its ready flow of supplies from the USSR and France, together with Iran's notorious shortage of aircraft spares. Following an appeal by UN Secretary-General Perez de Cuellar for a cease-fire over civilian-populated areas, this was accepted by both sides and the secretary-general now sought to widen it[8] into something that would give effect to the Security Council resolutions' call for an end to the fighting altogether.

This was unsuccessful. It became clear that Iran was still aiming at a battlefield victory, posing specific preconditions for an end to the war:[9] reparations and the removal of Saddam and his entire regime. Gradually the civilian area agreement began to crumble, each side citing violations by the other to UN observers.[10]

Other matters to which the secretary-general addressed himself were the plight of prisoners of war, a field in which the International Red Cross was also active,[11] and the use of chemical weapons. Here again, Iran was unhappy about the way in which the UN went about its work. After Iraq was condemned for its use of chemical weapons, President Khamenei was quoted as having wished the UN had gone on to take steps against both Iraq and the states which had provided it with the means to manufacture such weapons.[12] This thorough ventilation of the issue of chemical weapons did not bring the subject to a halt. Both sides were later reported either to have used them or to be planning to do so. In Iraq's case the most serious charges relate to Baghdad's efforts to subdue the insurgents in Kurdestan,[13] which gave rise to the chemical bombing of the village of Halabjah[14] in 1988 and the depopulation of Qalaat Dizah a year or so later. The charges of genocide that followed were denied by Iraq[15] but pressure accumulated for a UN investigation,[16] not least through the criticism of Turkish premier Ozal,[17] whose own country's relations with its Kurdish minority have enjoyed a chequered history over the years.

The NAM was another international body to become involved in the war: first, when its 1982 summit conference due to be held in Baghdad had had to be rescheduled. The Baghdad hosting had been arranged following a successful diplomatic campaign by President Saddam at the previous summit at Havana. At the Harare summit held in 1986, Iraq's adversaries were strongly represented, with notable appearances by Colonel Qaddafi and President Khamenei, whose main theme in his address was an unsuccessful demand for Iraq's expulsion from the movement.[18] The previous year the NAM had taken a direct initiative over the war when India's Prime Minister Rajiv Gandhi,[19] acting as NAM chairman, sent a peace mission to the Gulf.

Another movement to become involved in these efforts—one with a certain amount of membership in common with the NAM—was the rather younger[20] ICO. A body which Iran tended to distrust because of its GCC associations, its Good Offices Committee was unable to make headway and the plan to hold a summit in Kuwait in 1987 was criticized from the outset by Tehran,[21] which ultimately boycotted the occasion.[22] The GCC did not confine its efforts to end this unwelcome war to working through third parties like the ICO or the UN.[23] Indeed the sixth GCC summit held in Muscat in November 1985 took upon itself the task of ending the war the following year.[24] To such an end it had for some time been seeking to intensify contacts[25] with both Tehran and Baghdad. At the same time the GCC instituted collective security arrangements[26] whereby the six, while rejecting outside intervention, would defend themselves against any attack made upon any one of them.

Earlier that year, Saudi Arabia made a solo attempt[27] to establish lines of communication with Iran through a visit to Tehran by Foreign Minister Prince Saud al-Faisal. This maintained a trend of conciliation begun the previous year when Iran's Majles Speaker Rafsanjani was invited to meet King Fahd during his visit to Saudi Arabia for the annual pilgrimage[28] and Crown Prince Abdallah called for an end to the war and the submission of the two sides to 'just and honest mediation'.[29] An indication of the prospects of success of this and similar efforts was provided by the remarks of Ayatollah Khomeini.[30] President Saddam, he said, should 'be sent packing, either committing suicide or fleeing the country. We should submit neither to imposed peace nor imposed arbitration.' Hojjatoleslam Rafsanjani took it a step farther, affirming,[31] 'We will not discuss any plan which envisages the survival of the Baath Party in Iraq.' Saudi-Iranian relations were later to deteriorate following the disturbances caused by Iranian pilgrims at Mecca.

Other solo attempts were made by several countries. Syria, Algeria and Japan all tried to achieve outright peace, and Sudan tried to intercede on behalf of Egyptian prisoners of war in Iran as a special 'one-off' case. A senior Kuwaiti Foreign Ministry official, identifying a hitherto only peripheral actor[32] in the drama, characterized the role of the USSR in finding a solution as 'essential; the important cards of the Iran-Iraq war are now gathered in Moscow'. In the West the 'Group of Seven' (G7), at meetings in London and Venice, declared a concern to see the struggle brought to an end and freedom of navigation assured. A later postscript was the idea of investing the efforts of the UN Security Council with 'some enforcement mechanism'.[33]

12
The Final Exchanges

By early 1986 the land war was, if anything, going Iran's way. Ironically, it was to be Iranian offensives in the south that triggered the developments which brought the inland war close to an end, with Iraq holding the advantage. These were the fourth and fifth pushes in the Karbala series. One report[1] described observers in Baghdād as surprised that what was expected to be Iran's final offensive, taking in the Shatt island of Bovarian, began as a frontal assault against earthworks and water barriers east of Basrah where Iraq's defences were at their most formidable. The points at which fighting was fiercest were Shalamcheh[2] and Fish Lake, a reservoir[3] constructed by the Iraqi forces some years earlier as a barrier against anticipated attempts to take Basrah. Now Iraq claimed that the Iranian attackers were trapped in a killing zone by the lake.[4]

Once again there were reports of President Saddam at the battlefront. Tehran radio claimed that he had personally led one of the heaviest Iraqi counter-attacks which Iran claimed to have foiled at Fish Lake,[5] together with the wiping-out of three Iraqi brigades. Baghdad declared that its forces were mopping up west of Fish Lake, having 'purged' the lake of the first and second waves of Iranians to cross. Reporting of this kind had become familiar over the fighting in this area. One US daily had found the previous year that claims about Iran's original Fao offensive were contradictory[6] and that 'neither assertion can be confirmed'.

This was not quite the end, however; there were one or two loose ends to be tidied up. In March Iran announced that Karbala No 5 was finally over.[7] The human waves were at an end,[8] as Iraqi Defence Minister Khairallah had foretold in January.[9] Despite Hojjatoleslam Rafsanjani's assurances[10] that the object of Karbala No 5 was not Basrah, Iran had not quite abandoned all hope of taking it and made another attempt in April,[11] again approaching by way of Fish Lake and claiming numerous Iraqi losses. In September[12] Iran was again reported to be preparing a three-pronged offensive,[13] the central thrust aimed at Basrah. This was not the first attack of this kind; there had been another three-pronged assault the previous year in the Fao peninsula.[14] Neither bore much fruit; nor did a third near Basrah in December.[15]

The third had been expected for some time. Saddam had returned to the south, to keep an eye on developments,[16] as soon as an offensive was regarded as likely. With talk of the Iranians having mobilized half a million combat troops, Iraq made arrangements for a formidable assembly of Soviet helicopter gunships and other *matériel* to be waiting for them.

Then came a new development. In April and May 1988, having left the initiative over inland offensives for so many years to Iran, the Iraqis launched two swift attacks.[17] In the course of these they recovered both Shalamcheh and Fao itself,[18] apparently catching the Iranians quite unprepared, with many *baseej* killed or captured.[19] Up in the centre, Iran launched Karbala No 6 in the vicinity of Qasr-e-Shirin. The Iraqis later recovered the small Kurdish town of Haj Omran, captured by Iran some time before, and also the Iranian border town of Dehloran.[20]

Meanwhile, at UN headquarters in New York, the Gulf was commanding considerable attention. Successive Security Council resolutions[21] had failed to produce results. So had moves in the General Assembly and personal initiatives by the secretary-general earlier in the war. By July 1987, however, enough support had been generated, especially among the vital five permanent members of the Security Council, for the voting of Resolution 598 to be a relative formality. Moreover not only had Iraq, eager for an end to the war for some time, agreed to the principle of a cease-fire but Iran, initially sceptical and despite Ayatollah Khomeini's strong misgivings,[22] accepted the resolution, and also the proposal that the UN secretary-general visit Iran forthwith to study the practicalities of the cease-fire. After prolonged negotiations both at the UN and in the Gulf, the UN proposals were finally agreed,[23] although military operations continued into 1988 and fighting continued in the waters of the Gulf.[24]

UN Resolution 598 contained[25] provisions for:

(a) immediate cease-fire and withdrawal by both parties to their sides of the recognized international borders;
(b) dispatch of UN observers to supervise and verify adherence to (a);
(c) repatriation without delay of prisoners of war;
(d) both parties to co-operate with the UN secretary-general in his efforts at mediation in pursuit of a just and honourable settlement;
(e) other states to refrain from any act likely to lead to further escalation of the conflict; and
(f) the secretary-general to explore the scope for entrusting an impartial body with the task of inquiring into responsibility for the conflict.

The two belligerents having accepted this text in principle, Perez de Cuellar's task, involving a bout of shuttle diplomacy, was now to pin them down to an actual cease-fire on the ground. It emerged in the course of his travels that the way Iraq envisaged proceeding over (d) was not by contact with Iran through any third party, however prestigious, but by direct face-to-face discussions.[26]

This concept met resistance in Tehran, particularly as Iraq underlined its position with another offensive just after Iran's accepting the principle of a cease-fire.[27] Nevertheless by the beginning of September[28] the cease-fire had taken effect. One further Iraqi aim in continuing the land battle was no doubt to begin negotiations for a settlement from a position of maximum advantage.[29] Foreign Minister Tareq Aziz put forward proposals for practical steps towards a settlement, including the choice of either Tehran or Baghdad as the scene of the talks.

In the event, the two sides made heavy weather of them; it was in Geneva that the first unsuccessful attempt was made, followed by another, no more fruitful, in New York,[30] the two sides' disagreement extending to their respective versions of what occurred during their talks. By mid-1989[31] the subject had not moved forward very much, though there were hopeful indications for the repatriation of prisoners. It was not for another year, in the midst of another crisis, that a conclusion was finally reached. Baghdad gave way on the border issue in mid-August

1990, accepting Iran's conditions for a full peace treaty between the two sides—a move evidently intended to free the hands of its armed forces to contend with the opposition provoked in the Gulf by Iraq's invasion of Kuwait. It is hard to judge how far the change in personalities at the head of the Iranian regime following Khomeini's death in June 1989 simplified this process.

13
Desert Storm

The conclusion of a peace treaty between Iraq and Iran did not guarantee the tranquillity of the countries at the head of the Gulf. Once the original cease-fire had been put into effect in 1988, both parties found themselves with immense tasks of reconstruction to contend with and little, apart from their battered oil-export systems, with which to finance them. Baghdad had the additional burden of servicing its wartime loans from GCC oil producers, principally Saudi Arabia and Kuwait.

In Iraq's case, the export system had been given a boost with the completion late in 1989 of the IPSA-2 pipeline. The purpose was to bring Red Sea capacity to 1.6 million b/d, making 4.9 million in all, ignoring the Syrian line which seems unlikely ever to be reopened. The Fao area was repaired to the extent that it was possible for loading to be resumed at a restored Mina al-Bakr terminal.[1] By the end of 1989 Iraq had attained a production level of 2.9 million b/d, which may be compared with Iran's recovery for the same year to 2.8 million.[2] Out in the waterways of the Gulf, the days were calm, apart from the occasional strong north wind, or *shemaal*. In the Casualty Report page of *Lloyd's List*, once overflowing with incident caused by the war, the section headed 'Iran-Iraq Conflict' which even in 1988 was confined to 'carryover' items, progress reports on salvaged tankers, and so on was by 1989 almost empty.[3] Oil-revenue prospects seemed hopeful.

However, output is constrained by other factors such as the OPEC country quota, and oil revenues are determined by the amount of oil a producing country can sell and the unit price at which prevailing market conditions enable it to do so.

By 1990 Baghdad had come to perceive these calculations as imperilled by two GCC states, Kuwait and the UAE. Both countries had been over-producing against their OPEC quotas for some time, helping depress prices gradually and giving rise to familiar bouts of recrimination between them and other member states with higher absorptive capacities for revenues. In addition, Iraq viewed Kuwait as having gradually eaten away patches of Iraqi territory in the border area between them over the years[4] and also as draining reserves from part of Iraq's South Rumailah oilfield which straddled the frontier. Such were the charges levelled at Kuwait by President Saddam Hussein in his statements in mid-July 1990. He also threatened to repudiate the wartime debts and Tareq Aziz contributed further detail by valuing the oil Kuwait was said to have extracted at $2.4 billion.[5] Then in July Iraq concentrated troops at the border.[6] After a fruitless meeting set up at Jeddah in the hope of reconciling their differences and some display of nervousness by the oil market, Iraqi troops crossed into Kuwait on 2 August, occupying the capital and prompting the Amir, Shaikh Jaber, to seek refuge in Saudi Arabia. Later, the emirate was formally annexed and proclaimed a province of Iraq.

The international community's response took various forms. The UN Security Council was in session for a good deal of August, producing a series of resolutions condemning the invasion, imposing economic sanctions on Iraq, declaring Kuwait's annexation illegal, demanding free passage for diplomatic missions in Kuwait City, and so on.[7] IEA developed contingency plans; some Western states, concerned lest the crisis widen, took precautionary military steps; and certain Arab states also took steps. The Arab League was reported by the *San Francisco Chronicle* on 11 August 1980 as having voted 12:3, with 5 abstentions, to send forces to Saudi Arabia. Independent bodies like Amnesty International began to focus on the conditions in which the Iraqi takeover of Kuwait was being conducted. A similar interest was taken by the regional and international press in the course of their everyday activities.

Once the news of the invasion reached New York, the UN Security Council quickly adopted Resolution 660, condemning Iraq's action, and went on to vote a further 11 resolutions on related matters by the end of the year.

Independent commentators have not had much good to say about Iraq's occupation of Kuwait. A report published by Amnesty International in December 1990 accuses Iraq[8] of killing and torturing Kuwaitis and other atrocities. Looting also appears to have occurred on a spectacular scale.

The economic sanctions to which the international community resorted did not prove successful as a means of inducing Iraq to withdraw. They were no doubt intended to operate principally on exports of oil but Baghdad gained a propaganda advantage from drawing attention to their effect of depriving Iraq's population of imported food and medical supplies.

Towards the end of 1990 the conviction that only force would succeed in bringing about Iraq's departure was becoming widespread. On 29 November the Security Council voted Resolution 678, setting a departure deadline of 15 January and authorizing the use of 'all necessary means' for this result to be achieved in the UN's name 'and to restore international peace and security in the area'.

Having occupied Kuwait City on 2 August, Iraq pushed on southwards: in a press item two days later, 100,000 Iraqi troops were reported in a concentration near the Saudi frontier.[9] Drawing the conclusion that an all too possible sequel was a move across the frontier, and with the aim of pre-empting any such step, both the US[10] and Britain[11] decided to send forces to the area to protect Saudi Arabia. The US contribution was massive. Strong naval forces, including three aircraft carrier battle groups, headed for the Indian Ocean and other nearby areas; land-based airpower, F-15s and B-52s, was established in the Arabian peninsula; and the celebrated 82nd Airborne Division was also deployed to provide the beginnings of a ground force presence within what came to be spoken of as 'Operation Desert Shield'.[12] By the end of the year, these had all been solidly reinforced and also joined by French and a number of Arab contingents.

The 15 January deadline set by the UN was not met and Operation Desert Storm began. Hostilities between Iraq and the US with its European and Arab allies started on the evening of 16 January with an aerial assault on Iraqi forces in Kuwait and Iraq, using B-52s[13] based in Diego Garcia and, as time passed, various other points in and beyond the Middle East.[14] There were other Allied aircraft like the US F-117 Stealth bomber and RAF Tornado, and also the Tomahawk Cruise missiles fired by US Navy craft like the battleship USS *Missouri*.[15]

Iraq's response was threefold. First, Scud missiles were fired, both at Saudi Arabia (at targets in Dhahran and Riyadh)[16] and at non-belligerent Israel, in what was evidently a bid to reinforce the already solid Arab world popular support for Iraq's cause.

Second, an attempt was made to destroy Kuwait's oil-production system, allowing crude oil to flow into the sea from the Ahmadi Sea Island terminal, also setting the wells on fire.[17] The discharge of crude into the Gulf once again created a huge floating oil slick reported by Platt's as measuring 35 miles long by 10 miles wide (56 x 16 km), which endangered the seawater intake points of nearby desalination plants and also much of the resident bird population,[18] drawing charges of 'environmental terrorism'. US aircraft using the Smart laser-guided bomb attacked the pipeline manifold feeding the Sea Island and cut off the flow. Other wells continued to burn, causing atmospheric pollution on an epic scale. Specialist fire-fighters like the American Red Adair were called in to extinguish them; at one stage, US army tanks became involved and shelled some of the wells.

Baghdad's third response, announced on 29 January by the Iraqi High Command, was an attack on the small Saudi oil centre of Khafji. This was unsuccessful, but fighting continued there for some time. Some people saw this as an attempt to force the Allies' hand, provoking a premature initiation of the expected ground offensive which might then plunge into the Iraqis' prepared defence in depth.[19]

This game of bluff was played by the Allies as well. In the build-up to the ground offensive, preparations were made that were consistent with an amphibious assault on the Kuwait coast to draw the Iraqis' attention away from their right flank, the direction from which the main Allied thrust was eventually to come.

The second phase of Desert Storm was the long-awaited ground offensive. In the event, this proved to be a partial misnomer as the assault was accompanied by vigorous air support and preceded by the bombardment of missile sites in Kuwait[20] by the 16-inch guns of USS *Missouri*. Throughout this phase the solid figure of the operation's commander, General H. Norman Schwarzkopf,[21] was to be seen, busy both directing the course of the battle and fielding the inquiries of the media. The aim of the operation itself was twofold: first, to neutralize the Iraqi forces in both their offensive and defensive roles; and, second, to drive Iraq from Kuwaiti territory, while denying the Iraqi leadership the means of either reinforcing its forces there or threatening Kuwait again. All this was to be achieved by heading straight for Kuwait City and pinning down the Iraqis there; and then by moving from the west in substantial force, enveloping and destroying Iraq's elite reserve, the Republican Guard, with its T-72 tanks, described by French sources as 'the priority target'.[22] In fact, the Republican Guard was never destroyed at all, but survived to surround Kirkuk in August 1991.

The Allied forces had a reasonably clear idea of the dispositions of Baghdad's forces, while Saddam's intelligence was less complete. This was primarily due to the Allies' possession not only of AWACS aircraft but also the photographic coverage provided by earth-orbiting satellites.[23] During the Iran-Iraq war, Iraq appears occasionally to have been able to draw on a US coverage[24] to chart the movements of Iranian forces. This time, however, Baghdad had no reason to expect any such present.

Facing the Allies, mostly massed in their Hafr al-Baten assembly area in northern Saudi Arabia, was what Pentagon sources often described as the 'fourth largest army in the world'. As in the Iran-Iraq war, Iraq's forces were well dug in, with tanks sitting hull-down in prepared positions. Their total numbers were assessed as being of the order of 350–400,000 against the Allies' 500,000.[25]

Allied troops now punched deep into Kuwait and on into Iraq itself in a multi-pronged attack.[26] This began early in the morning of 24 February. Much of the attack was conducted by armour—US, British (including the 7th Armoured Brigade or 'Desert Rats') and French (including Foreign Legion troops). One of the prongs took the form of over 2,000 air assault troops of the US 101st Airborne Division, in what may well have been the largest helicopter-borne operation[27] ever mounted. During the whole of this offensive, many thousands of prisoners were taken;[28] the 101st established a base well inside Iraq and armour raced on to join them.[29] Some elements reached the Euphrates near Naseriyah. Other were even reported to have reached the outskirts of Baghdad itself,[30] though this was not confirmed.

The task of liberating Kuwait fell to a combination of US Marines, Saudi and Kuwaiti forces. Striking northwards up the coast from near Khafji,[31] accompanied also by some Qatari armour, they took it over the period 26–27 February. The Amir, however, did not return until mid-March.[32]

In a highly controversial episode, a large party of Iraqis, bolting homewards out of Kuwait in assorted vehicles, was blocked by Allied armour at a point called Mutla' Ridge, where Allied aircraft wiped out the entire group.[33] It has since emerged that several aircrew were reluctant to continue the slaughter.[34]

General Schwarzkopf, summing up the overall result on 27 February, described the rout of the Iraqi army as having been achieved by a 'flanking manoeuvre, trapping thousands of its troops'.[35]

The Allied air forces were meanwhile continuing their hunt for the sites from which Iraq was firing its Scud missiles at Saudi Arabia and Israel. The main problem was that the missile launchers were often mobile, moving from site to site. Although this hunt met with a certain amount of success, British SAS troops having been reported as active behind Iraqi lines pinpointing targets for US F-15 strike aircraft,[36] its results were not all positive. Another device employed to combat the Scuds was the US Patriot anti-missile missile, deployed in both target countries.[37] This again, though enjoying some success, could not be rated foolproof.

One important feature of the ground battle for which the Allied forces at all levels were profoundly grateful was the fact that the chemical warfare attacks for which they had prepared the protective equipment described in the *San Francisco Chronicle* of 19 February 1991 never materialized.

Many have asked, from the perspective of later events, why the Allied forces stopped when they were well within Iraq and failed to press home their advantage, take the capital and bring Iraq's leadership to account for its actions. This, however, was not what the Security Council had charged them to do[38] and their attention was focused on the liberation of Kuwait. Once this had been achieved, the fighting between Iraq and its adversaries came to an end and Iraq agreed to a meeting of military leaders to discuss initial cease-fire arrangements,[39] at which the Allies were represented by General Schwarzkopf and the principal Arab military figure, Prince Khaled ibn Sultan of Saudi Arabia. By early March, exchanges of prisoners under Red Cross supervision were being reported, to general relief.

Nevertheless, for President Bush, his Iraqi opposite number still represented a certain amount of unfinished business. As reported in *The Times* on 4 April, Bush called for forces within Iraq to unseat Saddam, anticipating a military coup mounted by disaffected generals. So far there has been no response of this sort; it fell to Iraq's minorities to make the attempt instead.

Revolt in Iraq

While Baghdad was preoccupied with securing a cease-fire, dissident forces quickly took advantage of a relaxation of internal vigilance: both in the north where Kurdish Peshmerga resumed their earlier insurgency, and in the south where Shiite counterparts rose in revolt against the regime in Basrah and other centres, with what Baghdad claimed was encouragement from Iran.[40] Opposition groups gathered in various countries. Mention has already been made of SAIRI, based in Tehran and headed by Hojjatoleslam Muhammad Baqer al-Hakim, and of the anti-Baghdad bloc fostered in Damascus. Two other organizations were now formed: the Iraqi Kurdestan Front, a grouping of several existing Iraqi parties with a largely Kurdish flavour; and the London-based Organization for Human Rights in Iraq, whose secretary-general was Dr Saheb al-Hakim, again a member of this well-known opposition family. In December 1990 the Syrian government had been as active as ever, encouraging the gathering in Damascus of a 'joint action committee' of familiar names—Hojjatoleslam Hakim, the Kurdish leaders Barzani and Talabani, and also Brigadier-General Naqib, in Syria since the early 1980s.

Baghdad's reaction to the twin revolt was severe. In the south, government forces bombarded the Holy Places of Najaf and Karbala,[41] later attacking Shiites who had taken refuge in the marshes north of Basrah.[42] They also paraded the senior Shiite divine, Grand Ayatollah al-Khoei, on television.[43]

In the north, after a wave of determined attacks against the regime, forcefully repelled, 1 million Kurds were reported to be in flight, trudging through the mountains, hoping to find sanctuary in Turkey and Iran.[44] This initially proved a mirage and despite feelers put out by Baghdad for the holding of talks to air their differences, resistance continued. In response, Baghdad maintained its attacks on Kurdish centres, sometimes making use of Sudanese mercenaries who were particularly hated by the Kurds. Dramatic media reporting of Kurdish suffering—'Halabjah is nothing compared with this'—so affected Western public opinion that there were widespread calls for the UN to intervene.[45]

The Security Council had in fact already been active, once the fighting was over. Resolution 686, adopted on 2 March, sought to reinforce the cease-fire. Then came Resolution 687, concerned with the restoration of Kuwaiti sovereignty and compensation claims, voted on 3 April. Another crucial step was the European Community's adoption of British Prime Minister Major's concept of establishing 'safe havens' inside Iraq, although there was still much to be done before a complete international consensus over the scheme could be established.[46] Even so, by the end of April 20,000 Kurdish refugees a day were returning to havens set up in northern Iraq by Western troops—not without some hesitation, as when Kurdish Peshmerga declared the haven at Zakho unsafe because there were still Iraqi police in the town.

A week or two earlier there were reports of a clash between the UN and the West, with harsh words being used of Secretary-General Perez de Cuellar. He was said to have been 'less than firm in his dealings with Baghdad' and to have generally failed to provide the leadership the Kurds' plight required,[47] prompting the decision to send US, British and French troops to protect them instead, despite concern, particularly in the US, not to keep troops in the region longer than necessary. Baghdad predictably objected to the concept of havens as interference in Iraq's internal affairs.

During the course of April, the feelers put out earlier by Baghdad were followed up; both Talabani and Barzani participated in talks with the Iraqi leadership in which the Kurds sought to secure a more workable version of the autonomy negotiated in 1970. This settlement was not made any more secure by the Allies' decision to withdraw from the area in June, a move strongly questioned by some of the forces involved.

Iraq's Weapons of Mass Destruction

The issue of nuclear and chemical weapons has been just as embattled as that of the Kurds, but this time the protagonists were the UN and the Iraqi government. The latter was notably reticent in opening its armouries to UN inspection in July and undergoing the anxiety of a further UN deadline the following month, provoking problems which have yet to be resolved. One comment attributed to the inspectors at a large chemical weapons plant at Muthanna near Baghdad was that this was the 'largest toxic waste dump in the world'.[48]

Just before the adoption of Resolution 689, concerned with the establishment of UNIKOM, the Iraq-Kuwait frontier Observation Mission, and a demilitarized zone, came the critically important 688, voted on 5 April. This addressed three issues which later emerged as vital.

First, it sought to strip Iraq of its weapons of mass destruction; next, it authorized international humanitarian relief inside Iraq for the Kurds;[49] and finally, it contemplated limited sales of oil to finance innocuous causes such as essential imports, the payment of reparations and the cost of maintaining a UN inspection team.

By the anniversary of the invasion of Kuwait, the groundwork had been laid for the introduction of far tighter UN controls over Iraq's activities in the advanced weapons field.[50] As for the Kurds, they have gone through another bad year and it remains to be seen whether they can be preserved from more of them. The omens are uncertain; the middle of August 1991 found government forces well established in the north again, with five brigades of Republican Guards reported as ringing Kirkuk.[51]

The War at Sea

This was in some ways just as dramatic as that of the 1980s, if only because of the very large warships—and their great number—being deployed in the area.

Mines were once again an ever-present hazard. In January 1991 there were sightings reported off Dubai and—as in 1983—off Qatar, while at Safaniyah in the Saudi offshore oilfield one of them struck a well platform.[52] Later on, some US Navy minesweepers off Kuwait were reported to have struck mines themselves.[53]

Although tankers did not come in for the attention they received in the 1980s, on 25 January Platt's carried an account of an attack on an Iraqi tanker by US aircraft. The vessel was said to have been intelligence gathering and the damage it sustained caused it to release oil and form another, though smaller, slick to float about the Gulf.

The French contribution to the Allied presence was confined to mine-hunters, the frigate *Dupleix* and other minor craft. The aircraft-carrier *Clemenceau*, though it set off for the Gulf[54] and its movements were followed initially by *Le Monde*, returned to Toulon in October after a 55-day voyage. According to one report, it never joined the other French craft in the Gulf at all.[55]

The War of Words

Baghdad's approach to this means of waging war was to begin by appealing to xenophobia. The form this took was to call for a *jihad*, or holy war, against the US forces and the overthrow of Saudi King Fahd and President Mubarak of Egypt.[56] A call was also made for a terrorist offensive. Although there was some limited response to the last of these, Iraq must have been disappointed by the overall results of the campaign.

Both Presidents Bush and Saddam Hussein had harsh words for one another in television broadcasts.[57] Saddam, a polished television performer, made considerable use of the medium for propaganda purposes, as when he conducted a conversation on camera with a British child caught up in the war who now found himself a 'guest' of the Iraqi regime.[58] The medium was also used for projecting news stories like the Allied bombing of the Ameriyah bunker near Baghdad.[59] This had been identified by the Allies as an Iraqi command and control centre, but in the event it proved to contain civilians, 300 of whom were reported killed. As the US television service CNN's transmissions were received[60] by satellite in Baghdad, it was a comparatively straightforward matter to assess the effectiveness of such activities.

Once again the newspaper cartoon was a favourite Iraqi medium for attacking the opposition, although in the US and other Western countries an independent press was more concerned with circulation than serving any political or military purpose. Another participant in this activity was the London-based paper *Sawt al-Kuwait*, in which the back-page cartoon was often an unflattering characterization of Saddam.[61] Another source of unwelcome attention came from agency reports of a rumour that the Iraqi president's family had been packed off to Algeria or Mauritania for the duration of the war.

One unusual communications medium employed on the battlefield was reported by *The Times*. Allied forces beamed 'heavy metal' rock music at Iraqi positions in Kuwait in the hope, we are told, of depressing their morale.[62]

Much of the energy of the press and television fraternity was devoted to battling for access to news material, sometimes involving conflicts within news 'pools' and sometimes between news-gatherers and the military, as when the Iraqi government expelled most journalists or the French agency AFP was reported to be bringing a lawsuit[63] against the Pentagon, which AFP accused of having excluded it from the pools.

The Fate of the Cities

Baghdad, as in the bunker incident at Ameriyah, came in for rough treatment when Allied airpower set out to paralyse the Iraqi nerve centre in January 1991. More than one correspondent filed a story describing the spectacle of a Cruise missile travelling up a Baghdad street at hotel window level.[64] Basrah also—as in the 1980s—received a severe pounding. It was described by one correspondent as a ghost city.[65]

Some of the smaller centres were also affected, as when an RAF Tornado making an attack west of Baghdad missed its target and hit the town of Fallujah, not far from Ameriyah, causing civilian casualties.[66] Nevertheless Allied airpower was probably the decisive element in determining the outcome of the war, with Iraq's aircraft so effectively neutralized in the opening stages as to be seldom a factor at all. (There were even reports of some of them being sent to bases in Iran well away from Allied bombing, much as had been done with civil aircraft in Amman in the 1980s.) It was also said that the Allies' freedom to employ satellites, both for intelligence and communications purposes, gave them a decided edge over their adversary.[67]

Throughout the war theatre, the centre that suffered the most was probably Kuwait City, both from being captured in August 1990 and retaken by the Allies the following February and from the Iraqi presence in between. Material devastation was on a spectacular scale, in addition to episodes such as the tale of 300 babies taken out of hospital incubators and left to die on the ground.[68] This was widely repeated, though doubt was later cast on the story's accuracy. During the final days, the occupiers were reported to have rounded up Kuwaiti civilians and carried them off to unknown fates in Iraq. Once the Iraqis had pulled out, Kuwaiti vigilantes began combing the city for any Iraqi soldiers still hiding in the ruins and for anyone known[69] or thought to have collaborated with the enemy, with Palestinians coming in for particular attention.[70]

Other legacies of the Iraqi presence in Kuwait included minefields and boobytraps. Its troubled aftermath now served to underline major divisions in Kuwaiti society, with calls for a more open political system, an improvement in the position of Palestinian and Iranian minorities, and so on. After the crisis was over, some Allied leaders began to wonder whether restoring Kuwaiti sovereignty had justified the effort involved—US Secretary of State Baker remarked that it was not 'the optimum type of regime'.[71]

Kuwait was not the only GCC society to be affected by the war. The presence of foreign armies—including their female personnel—in Saudi Arabia was not a development with which the more conservative elements were entirely comfortable, some of them describing the country with its Holy Places as being under the occupation of unbelievers. One notable sequel reported in the *Sunday Times* of 6 January[72] was a demonstration mounted by a number of Saudi women, not normally allowed to drive, who sent away their drivers and drove their cars around Riyadh.

Away to the north, Israeli cities had come under fire from Iraq's Scuds, with Tel Aviv and Haifa being the first targets. Washington meanwhile was acutely concerned to discourage retaliation by Israel[73] and its being thus drawn into the war. This prospect was seen as likely to destabilize the Alliance, because Arab members might well have had difficulty in explaining to their constituencies what amounted to an alliance with Israel against a fellow Arab state, Iraq.

In the West, the main impact of the war, apart from circles with relations serving in the forces, came through demonstrations against it. In London, Hyde Park was the scene of one such demonstration as Desert Storm approached, with the theme 'No blood for oil' prominent.[74] In the United States, the approach was often the same, but by the end of January 1991—possibly encouraged by the publicity given to the oil spill in the Gulf—attendance at peace rallies began to decline.[75] On 9 January a *Washington Post*-ABC poll had indicated that the US public was beginning to support President Bush's approach.

Friendly Fire

There were occasional episodes during Desert Storm in which—as in many wars—Allied forces were fired upon by their fellows, arousing an understandable combination of dismay and the concern to avoid any repetition. Probably the incident to receive the greatest publicity, at least in Britain, involved the shooting-up of two British Warrior armoured personnel carriers by an American A-10 Thunderbolt aircraft while all three were pursuing the retreating Iraqis, resulting in nine British infantrymen being killed.[76]

A similar incident involved the death of seven US Marines; the *San Francisco Chronicle* on 2 February carried a news story on a Pentagon investigation into the incident. In August 1991 the Pentagon revealed that of all US servicemen killed in Desert Storm, one quarter had fallen as a result of friendly fire.[77]

The Role of Diplomacy

Many attempts were made to bring the crisis to an end by means other than war. First came the Jeddah meeting at the beginning of August in 1990, followed by the fruitless meeting of US Secretary of State Baker and Iraqi Foreign Minister Tareq Aziz in Geneva in January 1991.[78] UN Secretary-General Perez de Cuellar visited Baghdad on 13 January as the bearer of a set of peace proposals from the UN.[79] The following month, Iran's President Rafsanjani offered to mediate, in the hope of bringing the war to an end.[80]

King Hussein of Jordan was from the outset, much as during the Iran-Iraq war, a supporter of Iraq. Early in 1991 he called on Arabs and Muslims to persuade the Allies to declare a cease-fire;[81] in July he saw both Tareq Aziz and President Mubarak in an attempt to intercede; and in the following month his efforts continued, in what was described as a 'last-ditch mediation effort'. His support for Baghdad was resented in US legislative circles and seemed likely to cost him any future US economic aid.[82]

Another supporter of Iraq that had to consider its position carefully was the PLO. Quite early on, its leader Yaser Arafat travelled to Baghdad in an evident gesture of solidarity. Over the years, however, the Palestinian mainstream had become accustomed to financial support from GCC states like Saudi Arabia and Kuwait and this support would now be called into question. Senior PLO figures like Hani al-Hassan nevertheless maintained the theme of support for Iraq.

One important move was that made by the USSR. A high-level envoy, Evgeny Primakov, visited Baghdad in February 1991,[83] but the sequel was probably not what he intended. On 15 February Iraq announced its intention of withdrawing from Kuwait, subject to certain conditions.[84] These were: the acceptance of a complete cease-fire; the withdrawal of all UN resolutions relating to matters other than the withdrawal itself; the abandonment by Israel of Palestine and all occupied territories; and the payment of reparations to Iraq. This announcement was, not surprisingly, met by the Allies with round condemnation.[85] Tareq Aziz travelled to

Moscow a few days later, perhaps with the aim of clearing the air,[86] where he encountered the Soviet leader Gorbachev.

Shortly after this,[87] Iraq decided to modify its position and was reported on 27 February to have accepted all the relevant UN resolutions unconditionally. This having been confirmed in writing, two things occurred: first, the Security Council's adoption of Resolution 686, to reinforce the cease-fire; and, second, the outbreak of the revolts in Iraq reviewed earlier in the chapter.

Despite all these efforts to prevent or limit warfare, there was a substantial toll in human life, though not as great as that of the Iran-Iraq conflict. Allied losses were described by General Schwarzkopf as 'remarkably light'. One European study[88] estimated Allied losses as 139 killed in action, while those of the Iraqis were assessed as up to 100,000. However, other figures have been advanced; the Pentagon's report on 'friendly fire' incidents indicates a figure for US deaths alone of 148. *The Times*[89] commented that both this figure and the accompanying data on troops wounded were higher than those previously issued.

In August 1991 the Security Council addressed the question of lifting the sanctions affecting sales of Iraqi crude oil, at last giving effect to what was envisaged in Resolution 688. On 16 August the UN adopted three resolutions concerning Iraq. Resolution 706 dealt with a partial export resumption, up to a total value of $1.6 billion over a period of six months, to enable Baghdad to import medical supplies and foodstuffs. The voting was 13 in favour, 1 against (Cuba) and 1 abstention (Yemen). The other resolutions were concerned with reparations, to which a maximum of 30 per cent of oil revenues were to be allocated, and, once again, with the control of weapons of mass destruction.[90]

There was no consistent pattern in Security Council voting over the crisis. During the initial month's activity in August 1990 which saw 10 resolutions adopted, only 4 were voted unanimously. Yemen abstained on 5 votes and Cuba opposed 2 and abstained on 3. Both countries were among the non-permanent membership of the Council, which sits for two years, the others for that particular two-year period being Canada, Colombia, the Ivory Coast, Ethiopia, Finland, Malaysia, Romania and Zaire. All five permanent members—the US, the USSR, Britain, France and China—voted for all 10 resolutions, though the USSR only figured as a co-sponsor of 3.

Oil Developments

Trouble in the Middle East is generally a signal to the oil markets for an increase in prices, implying a coming shortage of supply and a general willingness to bid higher for such oil as is on offer.

For the first six months of the crisis, prices were volatile, fluctuating between $20 and $40 per bbl.[91] They rose 10 per cent on 2 August,[92] and also moved up when the UN imposed sanctions, when Iraq refused the UN's call for withdrawal and when the UN set the 15 January deadline. After the initial airstrike, Far Eastern markets added $4 to the price of most grades of crude.[93] In November there was talk of general expectations in the Gulf of higher 1991 revenues (a 34 per cent increase was the figure advanced) compared with 1990.[94] (In the event, at the end of 1991 prices were weak, with Algeria's Nordine Ait Laoussine on Christmas Eve predicting worse developments still unless OPEC could see its way to reducing output.)

Important as the price of oil may be, the uncertainty over a possible physical shortage of supply was an equally compelling concern, with IEA active in formulating contingency plans for stock sharing in this event. This was the question of the hour for much of January 1991 and thoroughly documented by Platt's. EC stocks were reported as amounting to 100 days of consumption, implying a need for action. Helga Steeg, IEA's executive director, announced in January that IEA would go ahead with plans to make 2.5 million b/d available.[95] These were set in motion once fighting began and were kept in effect despite falls in prices.[96]

In March 1991 IEA issued its *Oil Market Report*,[97] with an update of its supply-demand balance for the first quarter of that year. The general dimensions of the issue had not changed much from those noted during the 1980s. World crude consumption outside the CPEs was forecast at 51.5 million b/d. Supply from non-OPEC sources was put at 28.6 million b/d, leaving a balance of 22.9 million b/d to be provided either from stocks or by OPEC. A figure published by Platt's for aggregate OPEC output for February[98] was also 22.9 million b/d, but care is obviously needed before any attempt can be made to relate the two. If we disaggregate the Platt's figure, Iraq and Kuwait are both 0 and the list is, as almost always, dominated by Saudi Arabia, this time with 8.3 million b/d.

As to the prospects of a resumption of oil exports by Iraq, Baghdad's State Oil Marketing Organization (SOMO) forecast this[99] for the beginning of the third quarter of 1991. In the case of Kuwait, however, prospects were more problematic.[100] A statement by the governor of the emirate's Central Bank in February noted that 650 wells were still on fire and put the time needed to resume output at 9 months. By August there was one favourable development. With Kuwait's Ahmadi refinery once again operational[101] at 130,000 b/d, the emirate was looking forward to being able to stop importing petroleum products at the end of that month. Crude oil exports from Kuwait were resumed in October 1991 when, oil industry circles revealed, Shell lifted a full ULCC cargo.

Over on the other side of the Gulf, Iran had its share of misfortune when an earthquake struck the region containing the key oilfield of Gach Saran.[102] On a more optimistic note, just as the 1990 crisis was developing, NIOC had been in the process of letting contracts for the reconstruction of the Kharg Island terminal.[103]

Some Consequences of the Crisis

One important effect of this conflict appears to have been to accentuate what many economists already saw as a serious drift on the part of the OECD world into recession.[104] By August 1991 the European Commission was reported as forecasting growth for the year at a discouraging 1.5 per cent.[105]

What the crisis certainly does not seem to have done, any more than the war between Iraq and Iran did, is to solve the more general problems of security in the Gulf, regional arms control and how to resolve conflict within OPEC. It is not enouraging to learn that Saudi Arabia and other states in the region are in the market for additions to their present arsenals of weaponry,[106] seeking in particular equipment used successfully during Desert Storm like the F-15 Eagle fighter, the M1 battle tank and the Rockeye cluster bomb.

As to OPEC, disagreement over adherence to individual country quotas is not new. The UAE, one of the states indicted by Iraq in 1990, had previously come in for criticism without any members finding it necessary to resort to violence to obtain redress. In 1991 the previous year's ceiling of 23.65 million b/d was rolled over for another year, but no country quotas were set.[107]

We have already noted the extent to which the crisis exposed unsolved social and political problems in both Iraq and Kuwait. One feature of life in the 1980s that the crisis did little to alter was the precarious existence of members of the upper reaches of the Iraqi system. On 6 December *Sawt al-Kuwait* drew attention to the continuing eclipse of the Iran-Iraq war hero Lieutenant-General Maher Abd al-Rashid; last heard of under house arrest in 1988, his fate was unknown in 1990.[108]

Postscript

August 1991 was not out before there was a new and dispiriting development. Reports came in of the landing of Iraqi military personnel on the uninhabited Kuwaiti island of Bubiyan. This aroused a wave of diplomatic protest and prompted the departure of a UNIKOM team to investigate British-Kuwaiti complaints. The island and its neighbour Warbah have been claimed by Iraq in the past and may well have been among Iraq's objectives in mounting the invasion a year earlier, although there was no strong evidence of this at the time.

Iraq's move certainly appears to have violated the cease-fire terms of UN Resolution 687.[109] For Perez de Cuellar, this incident came at a time when he was still closely involved with other business, notably engineering the release of hostages held in Lebanon and Israel. He described the incursion as the most dangerous development since the end of the war itself.[110]

His concern throughout the crisis had been to keep the principles of the UN Charter in the minds of all the parties involved; not only Article I, setting out the purpose of the UN, the maintenance of international peace and security, but Article VII, which excludes intervention in matters subject to any state's domestic jurisdiction. This aspect arose in particular over the setting up of safe havens for the Kurds.

Despite the uncertainties surrounding the future of the Kurds, the difficulties encountered in removing Iraq's weapons of mass destruction and the anxiety provoked by the Bubiyan incident, most of the endeavours embarked upon by the UN in the crisis appear to have turned out reasonably well.

It is true that a substantial body of opinion in various parts of the world would have preferred to see whether relying on sanctions for longer would have brought about Iraq's withdrawal. Even so, there seems to be a solid consensus, not to mention the Security Council majority which voted Resolution 678, that in the light of the failure of 'all efforts by the UN' up to then, the approach eventually adopted represented the most realistic course available. It is also clear that the UN as a whole has emerged from the crisis with greatly strengthened authority.

Notes

In the process of preparing this text, I have had recourse, in addition to the London daily and weekly press, to a wide selection of the regional and Western press and other publications. I indicate these in full recognition of the caution often needed in employing material of this kind, whether for purposes of reference or the initial analysis, and with the idea that the notes listed below should serve to provide an indication of publications in which the subject matter of the text can be explored more fully.

The general bibliography has not been set out separately but is instead to be found among the notes and source references listed below, and in some cases in the text itself. In addition, some useful background is available in:

Cambridge History of Iran (Cambridge University Press, Cambridge, 1968).

Cambridge History of Islam (Cambridge University Press, Cambridge, 1970).

Jane's Weapon Systems 1988–89 (Jane's Information Group, London, 1989).

Seton Lloyd, *The Archaeology of Mesopotamia* (Thames & Hudson, London, 1978).

Stephen H. Longrigg, *Oil in the Middle East* (Oxford University Press, London, 1954).

Abbreviations Used in the Notes

CSM *Christian Science Monitor* (Boston, Mass.)
CT *Chicago Tribune*
DT *Daily Telegraph* (London)
Econ. *The Economist* (London)
FAZ *Frankfurter Allgemeine Zeitung* (Frankfurt)
FT *Financial Times* (London)
Gdn *Guardian* (London)
IHT *International Herald Tribune* (New York and Paris)
Ind. *Independent* (London)
IPE *International Petroleum Encyclopedia* (Tulsa, Okla.)

KI	*Kayhan International* (Tehran)
LL	*Lloyd's List* (London)
LM	*Le Monde* (Paris)
MEED	*Middle East Economic Digest* (London)
MEES	*Middle East Economic Survey* (Nicosia)
NYT	*New York Times*
NZZ	*Neue Zürcher Zeitung* (Zurich)
Obs.	*Observer* (London)
PE	*Petroleum Economist* (London)
PON	*Platt's Oilgram News* (New York and Houston)
SFC	*San Francisco Chronicle*
ST	*Sunday Times* (London)
STel.	*Sunday Telegraph* (London)
SWB	*Summary of World Broadcasts* (BBC, London)
TASS	Soviet news agency
TT	*The Times* (London)
WP	*Washington Post*
WSJ	*Wall Street Journal* (New York)

Chapter 1

1. See Reuters, 23 Sept. 1980, with several counterparts (carrying the same date) in the international press, e.g. IHT: 'Iraq bombs airfields'; and FAZ: 'Bomben auf Flughafen und Stutzpunkte'.

2. There is detailed coverage of the incidents involved in this dispute in the monitor reports published by the BBC in its SWB over the period April-Sept. 1980. See also: (a) the regional and international press over this period, e.g. the Jeddah *Arab News* of 21 Sept. 1980, which carried an agency report from Associated Press datelined the previous day about 'contradictory reports on border fighting'; and (b) for a solid grounding in the Iranian perspective, Abdel Majid Trab Zemzemi, *La Guerre Irak-Iran* (Albatros, Paris, 1985), pp. 22–7. (In French.)

3. Known in Persian as the Arvand Rud, the combined flow of the rivers Tigris and Euphrates, in which 74 vessels were trapped in Sept. 1980—see MEES, 7 April 1986. Other reports give a higher figure, e.g. the 93 cited later by MEES on 13 Oct. 1986. See also LL, 2 May 1980.

4. CSM, 23 Sept. 1980. See also NYT, 30 Sept. 1980.

5. TT, 24 Sept. 1980.

6. *Arab News*, 24 Sept. 1980.

7. FT, 20 Sept. 1980. The diplomats were not identified.

8. Obs., 5 Oct. 1980, mentioned three divisions; TT, 19 Nov. 1980, reported nine.

9. TT, 24 Sept. 1980.

10. Gdn and FT, both of 24 Sept. 1980.

11. TT, 4 and 25 Oct. 1980; also NYT and Gdn, 26 Sept. 1980. See also LM, 27 Sept. 1980.

12. Obs., 5 and 26 Oct. 1980.

13. LM, 5 and 6 Oct. 1980; see also TT, 27 Sept. 1980 and DT, 29 Sept. 1980.

14. ST, 30 Nov. 1980.

15. TT, 25 Sept. 1980 and WSJ, 29 Sept. 1980.

16. A curious episode which followed Britain's decision to withdraw from the Gulf in the 1970s. See Fereydoun Hoveyda, *The Fall of the Shah* (Weidenfeld & Nicolson, London, 1979), pp. 53–5. See also TT, 2 Dec. 1971.

17. TT, 18 Sept. 1980 and *Arab News*, 20 Sept. 1980. See also *al-Jumhuriyah* (Baghdad), 22 Sept. 1980 and ch. 2, note 38, below.

18. TT, NYT and WP, all of 18 Sept. 1980.

19. TT, 6 Oct. 1980.

20. See first TT and Gdn, 26 Sept. 1980; also TT, 4 and 7 Oct. 1980.

21. TT, 30 Oct. 1980.

22. DT, 29 Sept. 1980.

23. TT, 11 Oct. 1980.

24. LM, 14 Oct. 1980.

25. Obs., 5 Oct. 1980. See also IHT, 1 March 1986.

26. ST, 28 March 1982.

27. CSM, 16 July 1982.

28. PE, Nov. 1980, p. 465 and Oct. 1981, p. 447. See also TT, 30 Sept. 1980. The combined pre-war capacity of these terminals was of the order of 2.5–3.0 million b/d.

Chapter 2

1. Ahmad ibn Yahya al-Baladuri, *Origins of the Islamic State* (Columbia University Press, New York, 1916), ch. 5, pp. 406–16 and *passim*.

2. *Al-Thawrah* (Baghdad), 1 Nov. 1980 and *al-Jumhuriyah* (Baghdad), 23 Sept. 1980. The theme was to recur both in these newspapers and in other Iraqi media as the war proceeded. See *Saddam Hussein on Gulf War* (Dar al-Ma'mun, Baghdad, 1988), pp. 62–4.

3. E.A. Wallis Budge and L.W. King (eds), *Annals of the Kings of Assyria* (Dept. of Antiquities, British Museum, London, 1902), p. 64 in which, in a cylinder inscription of Tiglath Pileser I, the monarch speaks of his campaigns: 'I traversed the mountains of Elama . . . I hewed my way with axes of bronze.' See also J.B. Pritchard, *Ancient Near Eastern Texts* (Princeton University Press, Princeton, N.J., 1955), pp. 75, 274 and 277, in the same spirit.

4. At Marathon, Salamis and Plataea. See George Rawlinson, *The History of Herodotus* (John Murray, London, 1858), vol. I, pp. 473–4.

5. Fereydoun Hoveyda, *The Fall of the Shah* (Weidenfeld & Nicolson, London, 1979), pp. 83–5.

6. See Roman Ghirshman, *Iran: Parthians and Sassanians* (Thames & Hudson, London, 1962).

7. See Muhammed ibn Jarir (Abu Jaafar) al-Tabari, *Chronique d'Abou Djafar* (Maisonneuve, Paris, 1958). (In French.)

8. Al-Baladuri, *Origins of the Islamic State*, pp. 409–16. See also al-Tabari, *Chronique d'Abou Djafar*, vol. 3, pp. 388 *et seq*.

9. Al-Baladuri, *Origins of the Islamic State*, pp. 409–16 and *passim*. See also al-Tabari, *Chronique d'Abou Djafar*, vol. 3, *passim*.

10. H. Lammens, 'al-Husein' in *Encyclopaedia of Islam* (Brill, Leiden, 1927), vol. II, p. 339.

11. See, for example, the *Sahih of al-Bukhari* (al-Azhar University, Cairo, 1936), a collection of the Prophet's authentic traditions, in which (p. 5) we find an account of the Prophet's revelation in the cave at Hira, where this experience occurred.

12. An example of the substitution of Arabic for Persian is the original text of al-Tabari's *Chronique*, by a Tabaristani Persian but written in Arabic. Something analogous, but to the reverse effect, occurred in the political history of the caliphate, during the Buyid period in Persia, when that dynasty's representatives in Baghdad, like Muizz al-Dawlah, dominated the caliphate to the extent of determining who was going to occupy it.

13. See Irene Melikoff-Sayar, *Abu Muslim* (Maisonneuve, Paris, 1962), *passim*. (In French.)

14. See Derek Hill and Oleg Grabar, *Islamic Architecture* (Faber, London, 1964).

15. A Shiite sect. See C. Huart, 'Ismailiya' in *Encyclopaedia of Islam*, vol. II, pp. 549–52. See also Bernard Lewis, *The Assassins* (Al Saqi Books, London, 1985).

16. Marco Polo, *The Travels* (Penguin Classics, London, 1958), pp. 113–62.

17. Wilfrid Blunt and Wim Swaan, *Ispahan, Perle de la Perse* (Albin Michel, Paris, 1967). (In French.)

18. C. Huart, 'Ithna Ashariya' in *Encyclopaedia of Islam*, vol. II, pp. 563–4.

19. Both the device employed and the concept of such a duty are reminiscent of the ruling by Ayatollah Khomeini in February 1989 following the publication of the controversial work *The Satanic Verses* (Viking, London, 1988) by the British author Salman Rushdie, whose condemnation to death in this manner became the occasion of a renewed estrangement between Britain and Iran. Note also TT, 18 Feb. 1989, suggesting the relevance of a power struggle in Tehran at the time.

20. Khaled al-Izzi, *The Shatt al-Arab Dispute* (Third World Publishing, London, 1981).

21. Gabriel Effendi Noradounghian, *Recueil d'Actes Internationaux de l'Empire Ottomane* (Cotillon, Paris, 1897) (in French), contains in the Contents the expression '1639–Traité de paix et de delimitation des frontières signé à Sehab près de Kasr e Shirin', but the text of this document is absent from the body of the work. It can, however, be found in J.C. Hurewitz, *Diplomacy in the Near and Middle East, 1535–1914* (Princeton University Press, Princeton, N.J., 1956), vol. I, pp. 21–3.

22. George N. Curzon, *Persia and the Persian Question* (Longmans Green, London, 1892); reprinted by Frank Cass, London, 1966.

23. For the text, see Sir Edward Hertslet, KCB, *Treaties between Great Britain and Persia, and between Persia and other Foreign Powers* (Butterworths, London, 1891), p. 169. It can also be found in al-Izzi, *Shatt al-Arab*, pp. 199–202.

24. See *British and Foreign State Papers*, vol. LVI, 1966, p. 648, containing a letter from Sir Stratford Canning, HM Ambassador in Constantinople, to Lord Palmerston in London, enclosing a copy of Canning's letter to Lieut.-Col. Williams, the British representative, containing travel instructions.

25. Lewis Hertslet, *Treaties between Turkey and Foreign Powers, 1535–1855* (Harrison & Sons, London, 1855).

26. Viscount Stratford de Redcliffe, *The Eastern Question* (John Murray, London, 1881), facing p. 216; the reference is to a line 'proposed by the Treaty of Berlin'. In the treaty itself, Article LX covers the cession to Persia of 'la ville et le territoire de Khotour'. See *Parliamentary Papers 1878*, vol. LXXXIII, pp. 675–709.

27. Readers may take their pick, once the tale has been told.

28. The Anglo-Iranian Oil Company (later British Petroleum) and Iraq Petroleum Company and its sister companies.

29. Mohamed Heikal, *The Return of the Ayatollah* (Andre Deutsch, London, 1981), p. 53. See also Peter Avery, *Modern Iran* (Ernest Benn, London, 1965), p. 54.

30. The concept whereby the frontier lies longitudinally along the centre line of the river bed. See al-Izzi, *Shatt al-Arab*, p. 101. It is also used in Articles II, XXXVI and XLV of the Treaty of Berlin.

31. Marco Polo, *Travels*, p. 35 (Bukhara) and p. 57 (Tabriz).

32. Ibn Battuta, *Travels in Asia and Africa 1325–1354* (Routledge & Kegan Paul, London, 1983), pp. 88 and 91.

33. Al-Izzi, *Shatt al-Arab*, pp. 33–4.

34. A letter from the British foreign secretary Sir Edward Grey to Sir Edward Goschen, the British ambassador in Berlin, in January 1913 contains a description of a conversation he had recently had with the Russian ambassador in London, in which Russia's interest in a settlement of the disputed Turco-Persian frontier was expressed. See G.P. Gooch and H.W.V. Temperley, *British Documents on the Origins of the War, 1898–1914* (HMSO, London, 1927), vol. X, part I, p. 424 (Doct 476). See also annotations on the Ordnance Survey 'Map of the Turco-Persian Frontier made by Russian and English Officers in the years 1849 to 1855' with references to 'British and Russian Notes of 5 August 1913 to the Ottoman Government'. See also note 36 in the present chapter.

35. For the text of this treaty see al-Izzi, *Shatt al-Arab*, pp. 221–4. See also *Foreign Office Papers* FO371/23255 in Public Record Office, London, in which a Baghdad chancery letter dated 16 February 1939 refers to a scheme to re-erect pillars on the Perso-Iraqi frontier in accordance with the Treaty of 4 July 1937, which itself draws on the 1913 Constantinople Protocol and the *procès-verbaux* of the Delimitation Commission of 1914.

36. The English-language Tehran daily *Kayhan International* (KI) of 15 March 1975 contains front-page coverage of the agreement, describing the land frontiers as having been set on the basis of the Constantinople Protocol of 1913 and the discussion records of the 'Border Demarcation Committee' of 1914. It also specifies the water borders as having been set along the estuary on the basis of the thalweg.

37. NYT and WP, 18 Sept. 1980. See also *al-Jumhuriyah* (Baghdad), 22 Sept. 1980, and SWB, 19 Sept. 1980, carrying President Saddam's speech on Baghdad radio on 17 Sept. 1980.

38. Again see ch. 1, note 2 above. See also WP, 13 May and 10 Sept. 1980; in addition, see NYT, 31 May 1980 and TT, 10 and 11 April 1980.

39. TT, 2 April 1980.

40. Ibid., 7 April 1980.

41. DT, 4 June and 7 Sept. 1979.

42. TT, 13 May 1980.

43. TT, 20 Sept. 1980; see also WP and TT, 21 Sept. 1980 and SWB, 12 April 1980.

44. SWB, 12 April 1980; see also DT, 15 May 1979 and ST, 13 April 1980.

Chapter 3

1. *Annuaire Statistique 1979–80* (United Nations, New York, 1981), p. 72.

2. The Mardom Party, set up as a counterpoise to the Melliyun (or Nationalists). See KI's front-page story, 4 Jan. 1975. See also Peter Avery, *Modern Iran* (Ernest Benn, London, 1965), p. 471.

3. The Rastakhiz (or Resurgence) Party. See Amir Taheri, *The Spirit of Allah* (Hutchinson, London, 1981), p. 211 and the entire front page of KI, 3 March 1975.

4. Mohamed Heikal, *The Return of the Ayatollah* (Andre Deutsch, London, 1981), pp. 94–6 and Taheri, *Spirit of Allah*, pp. 167–8. The preceding dynasty to the Pahlavis had not completely disappeared. TT, 9 May 1988, reported the death in London of Prince Hamid Qajar, son of the last Qajar Crown Prince, whose father had been deposed by Reza Shah in 1925.

5. Particularly agrarian reform, which affected some lands they traditionally administered. See Taheri, *Spirit of Allah*, pp. 124–5.

6. Heikal, *Return of the Ayatollah*, p. 94.

7. Fereydoun Hoveyda, *The Fall of the Shah* (Weidenfeld & Nicolson, London, 1979), p. 26 describes a riot in Isfahan and a mob running wild with cries of '*Shah haroomzadeh*' (The Shah's a bastard).

8. Ayatollah Ruhollah Khomeini, *Islam and Revolution* (Mizan Press, Berkeley, Calif., 1981), p. 254. The author cites an address he made in Tehran in February 1979: 'Muhammad Reza Pahlavi, the vile traitor, has departed. In his time only the cemeteries prospered.'

9. Ian Seymour, *OPEC, Instrument of Change* (Macmillan, London, 1980), pp. 123–6.

10. TT, 7 and 10 June 1963; see also NYT, 6 June 1963.

11. Hoveyda, *Fall of the Shah*, p. 19.

12. TT, 1 Nov. 1978. See also PON, 24 Oct. 1978 and PE, Feb. 1979.

13. PE, Dec. 1981, p. 548. See also TT, 7 Nov. 1978.

14. STel. and NYT, 24 Dec. 1978.

15. Brother of Fereydoun, the diplomat who wrote *The Fall of the Shah*.

16. Taheri, *Spirit of Allah*, p. 217.

17. Hoveyda, *Fall of the Shah*, pp. 115–16.

18. Khomeini, *Islam and Revolution*, *passim*.

19. Heikal, *Return of the Ayatollah*, p. 139.

20. Taheri, *Spirit of Allah*, p. 251.

21. DT, 17 Jan. 1979. Some of the atmosphere of this departure is conveyed in Christian Malar and Alain Rodier, *Reza Pahlavi* (Plon, Paris, 1986), pp. 71–3. (In French.) The last of the Shah's prime ministers, Shahpour Bakhtiar, was murdered in Paris in August 1991: see obituary in TT, 9 Aug. 1991 and LM, 17 Aug. 1991.

22. Heikal, *Return of the Ayatollah*, pp. 176–8.

23. See Edmund Wilson, *To the Finland Station* (Fontana, London, 1960), pp. 459–78.

24. ST, 8 June 1986 and TT, 9 June 1986.

25. Taheri, *Spirit of Allah*, p. 274.

26. Hoveyda, *Fall of the Shah*, p. 155.

27. Ibid., p. 42 and Heikal, *Return of the Ayatollah*, p. 183. See also TT, 5 Jan. 1980.

28. NYT, 27 April 1980, CSM, 28 April 1980 and WSJ, 29 April 1980.

29. Seymour, *OPEC*, p. 182.

30. UN, *Annuaire Statistique, 1979–80*, p. 72.

31. The revolution of 1958 which toppled the Hashemite monarchy and the two-stage coup of July 1968 which brought in the present Baathist regime.

32. Following the well-established traditions of British scholarship in this field, see: Sir Leonard Woolley et al., *Ur Excavations–Reports* (Oxford University Press, London, 1927) and *Selected Letters of Gertrude Bell* (Penguin, Harmondsworth, 1953), p. 377.

33. Hanna Batatu, *The Old Social Classes and Revolutionary Movements of Iraq* (Princeton University Press, Princeton, N.J., 1978), pp. 205–9.

34. Ibid., pp. 1073–110.

35. DT, 30 July and 10 Aug. 1979.

36. Batatu, *Old Social Classes*, p. 1092.

37. Barazan later fell from grace. See TT, 27 and 28 Oct. 1983.

38. DT, 15 Jan. 1979.

39. TT, 20 Sept. 1972.

40. Fereidun Fesharaki, *Revolution and Energy Policy in Iran* (Economist Intelligence Unit, London, 1980).

41. 'World Oil Production' in PE, Feb. 1985, p. 76.

42. 'Focus on OPEC Oil Companies, National Iranian Oil Company (NIOC)', *OPEC Bulletin* (OPEC, Vienna, April 1985), p. 22.

43. IPE, vol. 13, p. 183.

44. Winston S. Churchill, *The World Crisis* (Thornton & Butterworth, London, 1931), part I, ch. 5. The field, discovered in 1908, was Masjed-i-Suleiman, sometimes abbreviated to MIS. See also R.W. Ferrier, *History of The British Petroleum Company* (Cambridge University Press, Cambridge, 1982), vol. I, pp. 85–8 and Stephen H. Longrigg, *Oil in the Middle East* (Oxford University Press, London, 1954). pp. 17–22.

45. 'Focus on OPEC Oil Companies', p. 26. Again see IPE, p. 238. See also *Report for 1957 of Iran Oil Exploration and Production Company (IOEPC)* (Tehran, 1958), in British Library, London.

46. Fesharaki, *Revolution*, p. 182. See also PE, Nov. 1980, p. 444.

47. PE, June 1981, p. 232 and other issues.

48. IPE, p. 224. See also 'Focus on OPEC Oil Companies, Iraq National Oil Company (INOC)', *OPEC Bulletin* (OPEC, Vienna, March 1985), p. 29.

49. MEES, 30 July 1984. The new figures may betoken not merely a burst of exploration activity on the part of INOC but equally a concern to strengthen the case for a higher OPEC output quota.

50. IPE, pp. 186 and 239.

51. See Z.R. Beydoun, *The Middle East: Regional Geology and Petroleum Resources* (Scientific Press, Beaconsfield, UK, 1983), pp. 95–101.

52. TT, 25 Feb. and 1 March 1984. See also MEES, 27 Feb. 1984.

53. PE, June 1981, p. 232, and other issues.

54. Ibid.

55. MEES, 13 Oct. 1980 and 12 March 1984; see also PE, Nov. 1980, p. 465.

56. Econ., 20 Aug. 1988, p. 61.

57. APS Consultants in *Iraq Financial Report* (APS, Nicosia), 28 July 1986, p. 5.

58. *International Financial Statistics Yearbook 1988* (IMF, 1988), p. 419.

59. Econ., 20 Aug. 1988, p. 61.

60. The London-based International Institute for Strategic Studies (IISS), the US government's Arms Control and Disarmament Agency (ACDA), the Swedish-based Stockholm International Peace Research Institute (SIPRI) and the '*Jane*' publications (London).

61. Heikal, *Return of the Ayatollah*, pp. 106–7.

62. Front-page headline on KI, 1 Feb. 1975. See also Taheri, *Spirit of Allah*, p. 168; J.B. Kelly, *Arabia, the Gulf and the West* (Weidenfeld & Nicolson, London, 1980), p. 146; in addition, NYT, 5 Jan. 1975 and TT, 10 March 1975.

63. IISS, *The Military Balance 1980–81* (IISS, London, 1980), p. 42.

64. MEES, 23 July 1984.

65. Patrick Seale in Obs., 29 Sept. 1985.

66. Richard Johns in FT, 17 May 1984 and 29 Sept. 1985.

67. Widely reported since 1986. See particularly Obs., 30 Nov. 1986.

68. Obs., 2 Nov. 1980 and DT, 1 June 1984.

69. SWB, 8 Jan. and 21 March 1986.

70. MEES, 18 Aug. 1986 and 31 Aug. 1987. See also SWB, 3 Aug. 1985, TT, 13 June 1984 and WP, 18 Oct. 1987.

71. TT, 18 Aug. 1986.

72. Econ., 26 Dec. 1987 and FT, 1 June 1984.

73. See, in addition to various years of the IISS *Military Balance*, the Arabic-language *al-Taqrir* (London), 15–30 March 1988, which, in an article on 'The War of the Cities', reviewed the missile armoury of each side.

74. Gdn., 24 March 1984. See also Amos Perlmutter et al., *Ten Minutes over Baghdad* (Corgi Books, London, 1982), p. 78. This exploit was rehearsed for the Israelis by the Iranian air force in September 1980: see TT, 1 Oct. 1980. In LM, 30 Sept. 1980, it is clear that Israel, which disliked both belligerents, regarded Iraq as the worse of the two.

75. Gdn., 27 April 1984.

76. Ibid., 26 April 1984.

77. Obs., 6 March 1988.

78. TT, 1 April 1989.

Chapter 4

1. Samuel Brittan in FT, 12 Feb. 1979.

2. TT, 5 March and 21 May 1984.

3. STel., 20 Feb. 1984.

4. MEED, 1 June 1984.

5. TT, 27 Feb. and 24 May 1984. This is the Boeing E3 Sentry. See also NYT, 1 Oct. 1980, SFC, 9 Oct. 1980 and LM, 10 Oct. 1980. These aircraft were reported as being US-manned: see Obs., 17 June 1984.

6. MEES, 3 Oct. 1983.

7. Gdn, 8 June 1984.

8. DT, 6 June 1984.

9. MEES, 16 July 1984.

10. Ibid., 4 June 1984. Re. the spelling of Shaikh, etc: no systematic approach has been

employed in treating the transliteration of diphthongs, which mainly depends on that of the sources in which they occur.

11. TT, 7 June 1967 and NYT, 28 Sept. 1980.

12. See David Hirst in Gdn, 14 June 1984 and Jonathan Randal in Gdn, 4 June 1984.

13. *Al-Hawadeth* (London), 27 Dec. 1985.

14. TT, 18 Jan. and 24 May 1984. See also SWB, 8 and 17 Nov. 1984.

15. WSJ, 23 Sep. 1980.

16. TT, 24 May 1984.

17. Ibid., 17 May 1984.

18. MEES, 8 Dec. 1986.

19. Ibid., 11 Jan. 1988.

20. PE, July 1981, p. 283.

21. *OPEC Statistical Bulletin*, reproduced in MEES, 28 Nov. 1988.

22. DT, 30 May 1984.

23. FT, 23 May 1984. Further activity of this sort was reported in 1984 under the code-name Peninsula Shield II. See SWB, 11 Oct. 1984.

24. MEES, 5 Dec. 1983.

25. Gdn, 2 June 1984.

26. MEES, 16 Dec. 1985.

27. Ibid., 2 Jan. 1984.

28. TT, 18 June 1984.

29. SWB, 13 March 1986.

30. A reference to the (Christian) co-founder of the Baath Party. See ch. 3 above.

31. SWB, 25 March 1986.

32. MEES, 10 March 1986.

33. J.E. Hartshorn in MEES, 17 March 1986. See also Econ., 21 Sept. 1985.

34. MEES, 17 Feb. 1986.

35. SWB, 3 March 1986.

Chapter 5

1. The West Bank area of Jordan, occupied by Israel since the 1967 Arab-Israeli war and known to Jews as the biblical lands of Samaria and Judea; in addition there is the Gaza strip, similarly occupied in 1967.

2. Econ., 9 Nov. 1991 and *al-Ahram* (Cairo), 2 Nov. 1991.

3. B. Carra de Vaux, 'Djanna' in *Encyclopaedia of Islam* (Brill/Luzac, Leiden/London, 1913), vol I, pp. 1014–15.

4. TT, 15 Aug. 1988 and 5 May 1989. See also Econ., 26 Oct. 1985.

5. See, for example, TT, 28 Dec. 1991 and *The Middle East*, April 1990.

6. This is a personal assessment. I was living in Riyadh at the time.

7. Gamal Abdul Nasser, *Falsafat al-Thawrah* [Philosophy of the Revolution] (Dar al-Maaref, Cairo, 1955), p. 60. (In Arabic.)

8. Ian Seymour, *OPEC, Instrument of Change* (Macmillan, London, 1980), p. 16.

9. Ibid., p. 132.

10. William Eagleton Jnr, *The Kurdish Republic of 1946* (Oxford University Press, London, 1963).

11. TT, 10 March 1981.

12. Ibid., 13 Sept. 1980.
13. DT, 24 April 1979 and ST, 30 Dec. 1979.
14. Seymour, *OPEC*, p. 96 and TT, 31 July 1986.
15. Ibid., pp. 96–7.
16. Ibid., p. 129.
17. TT, 31 July 1986.

Chapter 6

1. This means 'The Eighth Imam', in the series of 12 imams venerated by the Ithna Ashariyah Shiites. As the war continued its course, the Iranian offensives were identified by SWB and less regularly in other reporting, though this improved towards the end of the war.

2. Between 29 Nov. 1981 and 30 Nov. 1982 came Tariq al-Qods (Road to Jerusalem), Matla' al-Fajr (Break of Dawn), Fath al-Mubin (The Clear Victory) and Bait al-Moqaddas (The Holy City).

3. These were Ramadhan al-Mubarak (Blessed Ramadhan), 13 July 1982; Muslim ibn Aqil (a figure in Shiite tradition, cousin of Husain ibn Ali), 1 Oct. 1982; and Muharram al-Haram (The Holy Month), 30 Oct. 1982.

4. 1983: Val Fajr 1 on 10 April; Val Fajr 2 on 20 July; Val Fajr 3 on 30 July; and Val Fajr 4 on 20 Oct. 1984: Val Fajr 5 on 15 Feb; and Val Fajr 6 on 22 Feb. The last in the series was Val Fajr 9 on 9 Feb. 1986.

5. IHT, 18 Jan. 1982.

6. ST, 4 March 1984 (under the headline 'Carnage at Gzaeil'), and Gdn, 9 March 1984.

7. Gavin Young in Obs., 10 June 1984.

8. DT, 23 March 1984. See also Gdn, 2 March 1984.

9. DT, 23 March 1984.

10. Major-General Sir Charles Townshend, K.C.B., *My Campaign in Mesopotamia* (Thornton & Butterworth, London, 1920), pp. 306 and 319.

11. See 'Iraq-Iran Map', published in Dehra Dun (India) by Surveyor-General of India, 1941–2.

12. Patrick Seale in Obs., 15 April 1984.

13. ST, 4 March 1984 and DT, 8 March 1984.

14. DT, 12 March 1984.

15. Gdn, 29 June 1984.

16. DT, 20 June 1984.

17. Gdn, 31 July 1984.

18. DT, 9 March 1984. Lieut.-General Maher Abd al-Rashid was the commander.

19. David Hirst in Gdn, 14 June 1984. One later inference drawn from the concentrations of armour and transporters of this sort was a desire to switch a whole armoured division from the south front to the centre, in the course of a night: see Econ., 28 Dec. 1987.

20. Gdn, 2 March 1984.

21. TT, 25 Feb. and 1 March 1984.

22. MEES, 8 July 1985.

23. FT, 10 March 1984. See also TT, 27 June 1988 and MEES, 25 June 1984.

24. DT, 12 March 1984.

25. MEES, 24 Feb. 1986. See also DT, 14 March 1984.

26. MEES, 5 March 1984. An earlier report in TT, 24 Feb. 1984 gave 22 villages liberated.

27. TT, 17 Feb. 1984 and Gdn, 23 Feb. 1984. See also MEES, 27 Feb. 1984.

28. SWB, 3 July 1986.

29. TT, 20 Sept. 1986.

30. Ibid., 27 July 1988.

31. Spread over five countries, the Kurds are thought to number some 18–20 million. See Gdn, 30 July 1984 and TT, 21 June 1988. See also William Eagleton, Jnr, *The Kurdish Republic of 1946* (Oxford University Press London, 1963), *passim*.

32. LM, 15 July 1989.

33. Obituary in TT, 4 Feb. 1987.

34. MEES, 15 Aug. 1983 and TT, 6 Aug. 1983. The spelling 'Qarzamand' also occurs in SWB during this period: see SWB, 30 July 1983.

35. *Middle East International*, 9 Nov. 1984 and TT, 7 April 1984.

36. Gdn, 16 March 1984.

37. MEES, 4 March 1985.

38. Ibid., 12 Sept. 1983. See also Econ., 18 June 1983.

39. MEES, 24 and 31 Oct. 1983.

40. FT, 15 Feb. 1984.

41. Abdel Majid Trab Zemzemi, *La Guerre Irak-Iran* (Albatros, Paris, 1985), p. 115. (In French.)

42. ST, 10 June 1984.

43. SWB, 6 and 7 March 1986.

44. Ibid., 25 March 1986.

45. TT, 2 Sept. 1986.

46. SWB, 26 Feb. 1986.

47. Ibid., 10 March 1986.

48. Ibid., 11 March 1986.

49. Ibid., 25 March 1986.

50. Obs., 7 Feb. 1988.

51. DT, 2 April 1988.

52. TT, 9 Sept. 1988. The Iranians had also shelled Halabjah; see SWB, 7 March 1986.

53. TT, 24 Sept. 1988.

54. Obs., 16 Oct. 1988.

55. SWB, 12 Feb. 1986 contains versions broadcast by both Baghdad and Tehran. See also TASS, 18 Feb. 1986.

56. MEI, 21 Feb. 1986.

57. MEED, 15 Feb. 1986.

58. SWB, 14 Feb. 1986.

59. Ibid., 20 Feb. 1986.

Chapter 7

1. LL *Supplement*, 21 May 1981. See also TT, 22 Feb. 1988, indicating an earlier incident, also covered in LL, 13 May 1981.

2. Reuter, citing the local navigational aid service, MENAS (Middle East Navigation Aids Service), 23 Sept. 1980.

3. The Centrally Planned Economies of the Comecon group, etc. Some energy economists employ the expression 'World Outside Communist Areas' (WOCA) to convey the same concept.

4. MEES, 23 April 1984. See also MEES, 19 March 1984. Another evaluation of the effect was the nearly 4 million b/d in PE, Nov. 1980, p. 465.

5. MEES, 4 June 1984.

6. Ibid., 16 July 1984.

7. Obs., 17 June 1984. The aircraft carrier was the USS *America*. Other carriers reported in the area were the USS *Midway* and the USS *Kittyhawk*. See also DT, 6 June 1984.

8. TT, 6 June 1984. See also TT, 8 June 1984.

9. MEES, 31 Dec. 1984.

10. Ibid., 20 Feb. 1984.

11. A personal recollection of the television coverage: I was in Baghdad at the time. See also MEES, 23 Aug. 1982.

12. DT, 1 June 1984.

13. Ibid., 2 July 1984.

14. MEES, 7 Nov. 1983. See also ST, 27 May 1984.

15. LL, 27 May 1987.

16. Ibid., 19 May 1984. See also FT and TT, both of 9 May 1984.

17. TT, 5 Sept. 1986.

18. Ibid., 16 March 1983.

19. MEES, 3 Nov. 1986. NIOC's options are reviewed in ch. 9 below.

20. TT, 27 July 1987.

21. MEES, 25 May 1987.

22. LL, 2 July 1987.

23. TT, 23 July 1983. See also Patrick Seale in Obs., 17 July 1983.

24. MEES, 7 Nov. 1983.

25. Ibid., 27 Aug. 1984.

26. Ibid., 16 May 1983.

27. Gdn, 26 March 1984. See also FT, 29 March 1984. Note as well the unconfirmed report in TT, 20 Oct. 1988, citing the Paris weekly *Le Canard Enchainé*, according to which French pilots were at times operating these aircraft. See also MEES, 18 Aug. 1986.

28. LL *Supplement*, 12 Dec. 1988, p. 17, though one fleet owner's unpublished analysis indicated 41.

29. MEES, 2 April 1984.

30. Obs., 1 April 1984.

31. FT, 10 May 1984. See also LL, 4 May 1984 and Gdn, 14 May 1984.

32. MEES, 21 May 1984.

33. Econ., 5 Dec. 1983 and DT, 19 May 1984. See also FT, 17 Feb. 1984 and TT, 29 May 1984.

34. SWB, 5 and 25 March 1986 and 14 Jan. 1987. See also MEES, 13 March 1986 and 1 June 1987; and FT, 21 May 1984.

35. MEED, 8 June 1984.

36. TT, 6 June 1984. See also Obs., 17 June 1984.
37. Obs., 3 June 1984.
38. FT, 18 May 1984.
39. LL, 11 July 1984.
40. MEES, 28 May 1984. See also FT, 26 May 1984.
41. MEES, 28 May 1984.
42. Ibid., 28 May 1984 and 31 Dec. 1984. See also FT, 26 May 1984.
43. MEES, 10 Dec. 1984; also 18 June, 22 Oct. and 29 Oct. 1984.
44. LL, 19 May 1984. See also DT, 30 May 1984.
45. MEES, 18 June 1984.
46. LL, 19 May 1984. See also DT, 30 May 1984.
47. TT, 5 Sept. 1986. See also Obs., 3 June 1984.
48. MEES, 4 June and 15 Oct. 1984. See also LL, 14 July 1984.
49. DT, 8 and 9 March 1984.
50. MEES, 13 Oct. 1983; also 31 Oct. and 7 Nov. 1983.
51. Ibid., 13 and 20 Aug. 1984.
52. NZZ, 16 Aug. 1985.
53. MEES, 13 and 20 Aug. 1984. See also TASS, 11 and 27 Dec. 1985.
54. LL, 5 Sept. 1986.
55. DT, 24 May 1984.
56. MEES, 19 Jan. 1987.
57. FT, 26 Oct. 1985. See also MEES, Oct. 1985.
58. SWB, 1 Jan. 1986.
59. MEES, 27 Oct. 1986 and 6 April 1987.
60. MEES, 20 and 27 April, 25 May and 8 June 1987.
61. TT, 30 March 1987.
62. Ibid., 24 July 1987. See also the American Middle West coverage in the *Daily Tribune* (Columbia, Mo.), 23 July 1987.
63. TT, 25 Jan. 1988.
64. MEES, 15 April 1985.
65. Ibid., 19 May 1986. See also TT, 3 April 1986.
66. TT, 26 Aug. 1986. See also MEES, 18 Aug. 1986.
67. Econ., 23 Aug. 1986. See also TT, 15 Aug. 1986 and MEES, 8 Dec. 1986. See also SWB, 6 Jan. 1987.
68. LL, 26 Nov. 1986. See also MEES, 1 Dec. 1986.
69. LL, 26 Nov. 1986. Larak was raided again the next year; see MEES, 12 Oct. 1987. See also MEES, 19 Jan. 1987.
70. MEES, 1 Dec. 1986.
71. LL, 26 Nov. 1986.
72. DT, 29 and 30 May 1984. See also Gdn, 6 June 1984.
73. TT, 1 Sept. 1986.
74. Ibid., 10 Jan. 1987.
75. LL, 29 Jan. 1988.
76. Obs., 15 May 1988. See also LL *Supplement*, 12 Dec. 1988, p. 30.
77. Kevin Shilleto of Internaft in MEES, 19 Oct. 1987.
78. LL, 9 Jan. 1988.

79. Ibid., 2 March 1984.
80. Ibid., 21 Feb. 1986.
81. Ibid., 2 June 1984.
82. MEED, 9 Nov. 1985.
83. Kevin Shilleto in MEES, 19 Oct. 1987.
84. LL *Supplement*, 12 Dec. 1988, p. 17.
85. MEES, 11 May 1987.
86. Ibid., 20 April 1987.
87. Obs., 29 March 1987 and MEES, 23 March 1987.
88. MEES, 6 Jan. 1986.
89. Ibid., 4 Nov. 1986. 'Dina' also occurs. See MEES, 10 Nov. 1986.
90. MEES, 10 Nov. 1986.
91. Ibid., 25 July 1986.
92. LL, 7 April 1987.
93. Ibid., 19 Sept. 1986.
94. MEES, 7 July 1986.
95. Ibid., 16 June 1986.
96. TT, 22 Feb. 1988.
97. LL *Supplement*, 12 Dec. 1988, p. 30.
98. TT, 26 Sept. 1987. See also MEES, 24 April 1988.
99. LL *Supplement* of 12 Dec. 1988, p. 29 and TT, 20 Oct. 1987. See also NZZ, 7 Aug. 1987.

Chapter 8

1. MEES, 13 Oct. 1980.
2. DT, 26 May 1984.
3. Gdn, 26 April 1984.
4. *Al-Taqrir* (London), 15–30 March 1988. See also Econ., 14 Sept. 1985.
5. MEES, 3 Oct. 1983. See also FAZ and Gdn, 24 Oct. 1983.
6. MEES, 30 Jan. 1984.
7. Ibid., 8 April 1986.
8. SWB, 22 Feb. 1986 and TT, 21 Feb. 1986.
9. MEES, 15 Feb. 1988.
10. Gdn, 11 May and 29 June 1984.
11. *Al-Thawrah* (Baghdad), 7 July 1982.
12. TT, 7 and 12 Aug. 1982. See also WP, 11 Aug. 1982.
13. TT, 29 June 1982. See also IHT, 3 Dec. 1982.
14. TT, 5 June 1982.
15. SWB, 7 and 16 Nov. 1984.
16. WSJ, 23 June 1983. See also SWB, 3 Jan. 1986, LM, 14 April and 14 Nov. 1982 and Gdn, 20 Nov. 1982.
17. TT, 19 and 22 Jan. 1988.
18. SWB, 6 Aug. 1985.
19. FT, 29 June 1982 and MEES, 10 Aug. 1987.
20. SWB, 18 March 1986 and WP, 30 July 1984. See also ch. 13.
21. Obs., 16 Oct. 1988.

22. TT, 8 May 1989.
23. Patrick Seale in Obs., 10 May 1987.
24. TT, 18, 20 and 22 Aug. 1986.
25. Ibid., 27 and 28 Dec. 1978. See also *Der Spiegel*, 16 Oct. 1988.
26. ST, 29 Jan. 1989.
27. TT, 26 March 1988.
28. Ibid., 2 Aug. 1984.
29. Obs., 24 May 1987.
30. Gdn, 22 June 1984.
31. Obs., 27 Sept. 1987.
32. MEES, 16 March 1987.
33. Gdn, 4 July 1984.
34. MEES, 24 Aug. and 5 Oct. 1987. See also *Der Spiegel*, 3 April 1989.
35. SWB, 27 Sept. 1980.
36. *Iran—Violation of Human Rights 1987–90* (Amnesty International, London, 1990).
37. TT, 13 May 1987 and 28 Jan. 1988. See also TT, 13 Dec. 1988 and *Iraq—Evidence of Torture* (Amnesty International, London, 1981).

Chapter 9

1. *OPEC Bulletin* (OPEC, Vienna), March 1985, p. 25. See also MEES, 13 Oct. 1980.
2. MEES, 2 Oct. 1983.
3. Ibid., 12 Dec. 1983.
4. SWB, 15 March 1986 and MEES, 24 Sept. 1984.
5. MEES, 10 Oct. 1983 and 6 April 1987.
6. Ibid., 29 April 1985 and 10 Feb. 1986.
7. Ibid., 30 Sept. and 7 Oct. 1985 and 10 Aug. 1987.
8. Ibid., 10 Aug. and 7 Sept. 1987.
9. Ibid., 30 May 1988.
10. Ibid., 3 Nov. 1986 and 30 May 1988.
11. Ibid., 8 Oct. 1984.
12. Ibid., 10 Aug. 1987.
13. For an assessment of proven reserves (7 billion bbl), see ibid., 27 Feb. 1984. See also FT, 25 Feb. 1984.
14. MEES, 1 Oct. 1984.
15. Ibid., 13 Oct. 1980.
16. Ibid., 21 Sept. 1987.
17. Ibid., 30 Jan. 1984 and 4 Nov. 1985.
18. Ibid., 4 Jan. 1985.
19. Ibid., 7 March 1988.
20. Ibid., 21 Sept. 1987.
21. Ibid., 7 March 1988.
22. NZZ, 8 Aug. 1986. See also MEES, 6 Oct. 1986.
23. PON, 4 Nov. 1980 and MEES, 6 Oct. 1986.
24. MEES, 6 March 1989.
25. Ibid., 30 Aug. 1982.

26. DT, 10 Dec. 1987.
27. MEES, 13 Oct. 1980.
28. Ibid., 6 Oct. 1986.
29. Ibid., 11 May 1981.
30. Ibid., 29 June 1981.
31. This was continuing in 1985; see MEES, 28 Jan. 1985.
32. Ibid., 22 Dec. 1980.
33. Ibid., 4 Nov. 1981.
34. In fact (see ibid., 29 March 1982) it was originally set at 18 million b/d, but after a unilateral Saudi cut of their quota from 7.5 to 7.0 million the ceiling was altered to match. The initial quotas allotted to Iraq and Iran were both 1.2 million. Bringing the picture up to date, the ceiling for the first half of 1989 was 18.5 million, but it went through a turbulent period and was raised at that summer's Vienna conference to 19.5, with indications it might later go still higher. See TT, 8 June 1989.
35. The share settled on was 18 million b/d; see MEED, 1 Feb. 1986.
36. MEES, 27 May 1985; see also Econ., 21 Sept. 1985 and Robert Mabro and Nordine Ait Laoussine, 'We are Worried', MEES, 27 Jan. 1986.
37. MEES, 17 March 1986. Another Saudi measure reported during this period (MEED, 16 Feb. 1984) was the negotiation of a Saudi-British deal for the supply of aircraft, notably the Tornado fighter-bomber, to be paid for in crude oil.
38. See Nordine Ait Laoussine and John C. Gault, 'The 1986 Oil Price War; an Economic Fiasco', in MEES, 6 Oct. 1986, describing a price fall of $20 per bbl and an overall revenue loss for OPEC of $50 billion.
39. MEES, 10 Feb. 1986.
40. Ibid., 5 Oct. 1987. See also NZZ, 8 Aug. 1986.
41. TT, 6 Aug. 1986.
42. MEES, 31 Oct. 1988. See also *al-Qabas* (Kuwait), 21 Sept. 1988, FT, 8 Aug. 1986 and MEES, 7 Dec. 1987.
43. Again see MEES, 31 Oct. 1988; also that of 7 Dec. 1987.
44. Ibid., 10 Oct. 1988.
45. Ibid., 14 Sept. 1987.
46. Ibid., 21 March 1983.
47. FT, 8 Aug. 1986 and MEES, 11 Aug. 1986.
48. PE, Aug. 1980, p. 329.
49. Shell International Petroleum Co. Ltd, in *Minutes of Evidence before the House of Commons Foreign Affairs Committee* (HMSO, London, 24 Feb. 1988), p. 29.
50. A 1980 comment by PON indicated a global figure of 1.8 billion bbl over minimum working levels.
51. This continues. The OMR for March 1989, of which extracts appear in MEES, 13 March 1989, indicates a figure of over 50 million b/d for 1989 for world demand excluding CPEs. So does the Merrill Lynch publication *Energy Monthly*, cited in MEES, 13 March 1989.
52. MEES, 21 Sept. 1987.
53. Ibid., 11 Feb. 1985.
54. Ibid., 30 April 1984. See also SWB, 6 Oct. 1984.
55. Reuters, 17 Feb. 1984.

56. MEES, 14 Jan. 1985.

57. Ibid., 21 Jan. 1985.

58. Ibid., 30 Nov. 1987 and 26 Sept. 1988.

59. Ibid., 24 Feb. 1986.

60. *World Financial Markets* (Morgan Guaranty Trust, New York), Oct.-Nov. 1986, p. 2.

61. MEES, 30 Nov. 1987.

62. Ibid., 18 May 1987.

Chapter 10

1. TT and SWB, 27 Sept. 1980.
2. SWB, 3 March 1986.
3. Ibid., 22 Feb. 1986.
4. Ibid., 15 Jan. 1987.
5. Ibid., 17 July 1987.
6. Ibid., 8 Jan. 1986.
7. Ibid., 20 Feb. 1986.
8. TT, 27 Feb. 1987.
9. SWB, 26 Sept. 1980.
10. Ibid. See ch. 3, note 28.

Chapter 11

1. FAZ, 18 Oct. 1983.
2. Gdn, 16 and 17 Feb. 1984. See also DT, 21 Feb. and 6 March 1984.
3. TT, 2 June 1984.
4. Gdn, 2 June 1984. See also TT, 31 May 1984 and FT, 2 June 1984.
5. TT, 1 Nov. 1983.
6. ST, 23 Oct. 1983.
7. Gdn, 22 Feb. 1984.
8. TT, 14 June 1984.
9. Gdn, 30 June 1984. A 1980 report in TT had indicated Basrah as a possible element in reparations to Iran.
10. TT, 22 June 1984
11. Gdn, 16 and 17 Feb. 1984. See also FT, 21 Feb. and 6 March 1984.
12. SWB, 25 March 1986.
13. Obs., 9 April 1989.
14. TT, 22 Sept. 1988. See also Helga Graham in Obs., 27 March 1988, and TT, 13 Sept. 1988. SWB, March 1986, covers Iran's attacks on the area in the course of Val Fajr No 9. For Qalaat Dizah, see Obs., 9 April 1989.
15. Obs., 18 Sept. 1988.
16. TT, 9 Sept. 1988.
17. Ibid.
18. Ibid., 3 and 5 Sept. 1986.
19. Ibid., 22 March 1985.
20. The ICO was founded in 1969; the NAM goes back to Nehru, Tito and Nasser.

21. SWB, 6, 10 and 15 Jan. 1987. See also MEES, 12 Jan. 1987 and *al-Sharq al-Awsat* (London), 2 and 3 May 1985.

22. Obs., 1 Feb. 1987.

23. TT, 22 May 1984. See also MEES, 21 May 1984.

24. *Al-Hawadeth* (London), 25 Oct. and 8 Nov. 1985.

25. SWB, 20 Dec. 1985.

26. MEES, 25 June 1984.

27. *Al-Sharq al-Awsat*, 19 May 1985. The later deterioration in relations between Iran and Saudi Arabia in 1987, due to events at that year's pilgrimage, is covered in TT, 1 Aug. 1987 and *Der Spiegel*, 10 Aug. 1987.

28. FT, 17 July 1984.

29. TT, 25 Aug. 1986.

30. Gdn, 30 June 1984.

31. MEES, 11 Feb. 1985.

32. SWB, 20 Dec. 1985.

33. MEES, 8 June 1987.

Chapter 12

1. Obs., 18 Jan. 1987.
2. FT, 10 Jan. 1987.
3. TT, 8 April 1987.
4. Ibid., 13 Jan. 1987.
5. SWB, 13 Jan. 1987. See also Obs., 18 Jan. 1987.
6. NYT, 26 Jan. 1986.
7. MEES, 2 March 1987.
8. Obs., 1 March 1987.
9. SWB, 6 Jan. 1987.
10. Ibid., 19 Jan. 1987.
11. TT, 8 April 1987.
12. Ind., 21 Dec. 1987.
13. Obs., 20 Sept. 1987.
14. MEES, 5 May 1986.
15. Ind., 21 Dec. 1987.
16. TT, 3 Dec. 1987.
17. MEES, 30 May 1988.
18. TT, 18 July 1988.
19. Obs., 29 May 1988.
20. FT, 15 July 1988.
21. UN Security Council Resolutions 479 of September 1980, 514 of July 1982, 552 of June 1984 and 582 of February 1986; from Security Council Proceedings for the years in question, New York, in following years; available at United Nations Information Centre, London as well as UN Headquarters in New York and Geneva. See also *Saddam Hussein on Gulf War* (Dar al-Ma'mun, Baghdad, 1988), pp. 27–9.
22. For Iraq's agreement, see TT, 19 June 1987; for Iran's initial reaction, see TT, 21 July 1987. The Imam was reported as finding the task more deadly than poison: see TT, 20 July 1988 and Obs., 24 July 1988.

23. FT and LM, 19 July 1988 and ST, 24 July 1988. See also Econ., 8 Aug. 1988.
24. TT, 1, 2, 5 and 10 Sept. and 23 Oct. 1988.
25. MEES, 25 July 1988.
26. TT, 28 July 1988.
27. Ibid., 25 July 1988.
28. Ibid., 2 Sept. 1988.
29. Ibid., 25 Sept. 1988.
30. MEES, 17 Oct. 1988.
31. Ibid., 1 May 1989.

Chapter 13

1. MEES, 19 June 1989.
2. *Oil & Gas Journal*, 31 Dec. 1990.
3. LL, 21 April 1989.
4. TT and LM, 20 July 1990. See also MEES, 25 June 1990.
5. FT, 19 July 1990.
6. WP, 4 Aug. 1990.
7. Numbers 660–665 in *UN Security Council Resolutions Relating to the Crisis in the Gulf* (UN Dept of Public Information, New York, 1991).
8. SFC, 19 Dec. 1990.
9. WP, 4 Aug. 1990.
10. ST, 5 Aug. 1990 and WP, 9 Aug. 1990. See also General Carl E. Vuono, US Army Chief of Staff, in *Foreign Affairs* (New York), spring 1991, pp. 49–68.
11. TT, 9 Aug. 1990 and Gdn, 27 Nov. 1990.
12. TT, 21 Aug. 1990.
13. SFC, 17 Jan. 1991.
14. CT, 2 Feb. 1991. Mentioned at different times have been Incirlik in Turkey, Thumrait in Oman, and Fairford (Glos.) in the UK.
15. ST, 27 Jan. 1991. See also *Sawt al-Kuwait* (London), 8 Feb. 1991 and *al-Hayat* (London), 19 Jan. 1991.
16. ST, 20 Jan. 1991. See also *al-Hayat*, 19 Jan. 1991 and Peter Hillmore in Obs., 17 Feb. 1991.
17. When a UN mission visited Kuwait within the framework of UNIKOM (created pursuant to UN Resolution 689) to assess the damage to the country's infrastructure from August 1990 to February 1991, it reported that of 1,330 active wells, 700 were on fire.
18. ST, 27 Jan. 1991 carried a photograph of a stricken cormorant on its front page. See also PON, 29 Jan. 1991 and Obs., 24 March 1991.
19. Gdn, 30 Jan. 1991. See also TT, 15 March 1991.
20. Gdn, 25 Feb. 1991. Among other craft involved was the USS *Wisconsin*; see photograph in *al-Hayat*, 19 Jan. 1991.
21. ST, 23 June 1991 and NYT, 25 Feb. 1991.
22. LM, 8 Feb. 1991.
23. Gdn, 18 Sept. 1990.
24. WP, 15 Dec. 1986.
25. Edward Foster and Rosemary Hollis, *War in the Gulf: Sovereignty, Oil and Security*, Whitehall Paper no. 8 (Royal United Services Institute for Defence Studies, London,

1991), pp. 162–74. Some other figures can occasionally be seen for the Iraqi side, up to a total strength of roughly 1 million, which probably includes People's Army units.

26. NYT and TT, 24 Feb. 1991. See also TT and FT, 25 Feb. 1991 and LM, 26 Feb. 1991.

27. NYT, 25 Feb. 1991.

28. CT, 25 Feb. 1991. See also WP, 24 and 25 Feb. 1991 and LM, 26 Feb. 1991.

29. NYT, TT and Gdn, 26 Feb. 1991. See also FT and WP, 25 Feb. 1991.

30. Gdn, 25 Feb. 1991. A more sober account in WP, 28 Feb. 1991, put the 'Screaming Eagles' of the 101st less than 100 miles (160 km) from the Iraqi capital on the 27th.

31. SFC, 28 Feb. 1991.

32. NYT and CT, 15 March 1991.

33. TT, 2 and 27 March 1991.

34. British Foreign Secretary Douglas Hurd in TT, 2 Aug. 1991.

35. SFC, 28 Feb. 1991.

36. Obs., 2 June 1991. The SAS had been reported in action in Kuwait earlier in the crisis, when they made off with an Iraqi SA missile: see ST, 13 Jan. 1991.

37. LM, 13 Feb. 1991.

38. Again see Douglas Hurd in TT, 2 Aug. 1991. But see also former Labour Attorney-General Lord Shawcross in Obs., 7 April 1991.

39. TT, 1 March 1991. See also TT, 22 June 1991 on the release of one particular prisoner.

40. SFC, 21 March 1991 and NYT, 9 March 1991. See also Obs., 10 March 1991.

41. Obs., 10 March 1991.

42. Ibid., 9 June 1991. See also Ind., 8 July 1991.

43. Obs., 10 and 24 March 1991.

44. TT, 5 April 1991.

45. Julie Flint in Obs., 7 April 1991. See also ST, 14 April 1991 and TT, 4 April 1991.

46. TT, 9, 18 and 30 April 1991. See also ST, 28 April 1991.

47. TT and Ind., 18 April 1991.

48. TT, 14 June 1991; also 4, 19 and 25 July and 1 Aug. 1991.

49. Again see Douglas Hurd in TT, 2 Aug. 1991.

50. Ibid., 3 Aug. 1991.

51. LL, 16 Aug. 1991.

52. PON, 22 Jan. 1991.

53. NYT, 19 Feb. 1991.

54. LM, 13, 14, 27 and 29 Aug. 1991.

55. ST, 7 Oct. 1990.

56. FT, 11 Aug. and 6 Sept. 1990. See also WP, 21 Jan. 1991 and TT, 5 Feb. 1991.

57. TT, 12 July 1991.

58. This episode appears in ITN's 1991 video film *The Gulf War*, assembled from the network's news coverage. See TT, 11 Dec. 1990, dealing with the child's homecoming.

59. Julie Flint in Obs., 17 Feb. 1991. See also ST, 17 Feb. 1991, Gdn, 24 Feb. 1991 and PON, 14 Feb. 1991. Not all the Allied air forces' attacks had this kind of outcome; see *al-Hayat*, 19 Jan. 1991, in which the front-page photograph shows smoke rising from the Ministry of Defence in Baghdad.

60. ST, 20 Jan. 1991.

61. During the early months of the crisis, examples were *Sawt al-Kuwait*'s issues of 21 Nov. and 14 Dec. 1990. The following year this continued; for example, see the 31 July 1991 issue. For an example of coverage of movements of the presidential family, see *al-Hayat*, 19 Jan. 1991.

62. TT, 22 Feb. 1991.

63. LM, 4 Feb. 1991.

64. The BBC's John Simpson in ST, 20 Jan. 1991. See also the ST reporter Marie Colvin in ST, 27 Jan. 1991. In TT, 8 Feb. 1991 there is a photograph showing the effect of the landing of a Cruise missile on the Baghdad suburb of Karradah.

65. LM, 25 Feb. 1991.

66. ST, 17 Feb. 1991. See also FT, 18 Feb. 1991.

67. Sir Peter Anson and Dennis Cummings, 'The First Space War', *Royal United Services Institute (RUSI) Journal* (winter 1991), p. 45.

68. SFC, 10 Dec. 1990.

69. Obs., 10 May 1991.

70. SFC, 20 March 1991. See also TT, 6, 8 and 26 March, 8 April, 25 May, 14 June and 4 July 1991, Obs., 28 July 1991 and *Sawt al-Kuwait*, 24 July 1991.

71. SFC, 23 Jan. 1991. See also TT, 14 June 1991 and Obs., 7 April 1991.

72. Also covered in NYT, 7 and 8 Nov. 1990.

73. WP, 18 Jan. 1991.

74. ST, 13 Jan. 1991.

75. Ibid., 27 Jan. 1991.

76. Again see ITN's video film *The Gulf War*.

77. TT, 14 Aug. 1991.

78. ST, 13 Jan. 1991.

79. Gdn, 14 Jan. 1991.

80. FT, 5 Feb. 1991.

81. WP, 7 Feb. 1991. See also Obs., 17 Feb. 1991.

82. WP and SFC, 21 March 1991.

83. PON, 14 Feb. 1991 and WP, 12 and 14 Feb. 1991. See also TT, 13 Feb. 1991.

84. Julie Flint in Obs., 17 Feb. 1991. See also WP, 16 Feb 1991.

85. Obs., 17 Feb. 1991.

86. FT, 18 Feb. 1991.

87. TT, 23 and 28 Feb. and 1 March 1991.

88. Foster and Hollis, *War in the Gulf*.

89. TT, 14 Aug. 1991.

90. LM, FAZ and Econ., 17 Aug. 1991.

91. PE, Feb. 1991, pp. 4–7.

92. Gdn, 3 Aug. 1990. See also FT and PON, 2 Aug. 1991.

93. PE, Feb. 1991, pp. 4–7.

94. LL, 29 Nov. 1990.

95. PON, 14 Jan. 1991. See also Obs., 27 Jan. 1991.

96. PON, 29 Jan. 1991.

97. Ibid., 6 March 1991.

98. Ibid., 1 April 1991.

99. Ibid., 26 April 1991.

100. Ibid., 28 Feb. 1991.
101. LL, 19 Aug. 1991.
102. PON, 15 Feb. 1991.
103. MEES, 25 June 1990.
104. PE, Feb. 1991, p. 4.
105. FT, 14 Aug. 1991.
106. Sir Anthony Parsons in TT, 13 March 1991 and Obs., 28 April 1991.
107. PON, 27 Nov. 1991 and MEES, 2 Dec. 1991.
108. *Sawt al-Kuwait*, 6 Dec. 1990.
109. TT, 30 Aug. 1991 and DT, 31 Aug. 1991. See also Gdn and FT, 30 Aug. 1991.
110. *Sawt al-Kuwait*, 31 Aug. 1991.

Index